WOOD CHEMISTRY
Fundamentals and Applications

WOOD CHEMISTRY
Fundamentals and Applications

EERO SJÖSTRÖM

Laboratory of Wood Chemistry
Forest Products Department
Helsinki University of Technology
Espoo, Finland

1981

ACADEMIC PRESS

A Subsidiary of Harcourt Brace Jovanovich, Publishers

New York   London   Toronto   Sydney   San Francisco

ACADEMIC PRESS, INC.
111 Fifth Avenue, New York, New York 10003

United Kingdom Edition published by
ACADEMIC PRESS, INC. (LONDON) LTD.
24/28 Oval Road, London NW1 7DX

Library of Congress Cataloging in Publication Data

Sjöström, Eero, Date.
  Wood chemistry.

  Revised and expanded translation of: Puukemia.
  Includes bibliographies and index.
  1. Wood--Chemistry. I. Title.
TS932.S5813   674'.134   81-3614
ISBN 0-12-647480-X   AACR2

PRINTED IN THE UNITED STATES OF AMERICA

81 82 83 84   9 8 7 6 5 4 3 2 1

CONTENTS

9 Cellulose Derivatives

10 Chemicals from Wood and By-Products after Pulping

PREFACE

Despite the rapid development of the disciplines allied with wood chemistry, nearly two decades have elapsed since the last English-language book devoted specifically to this subject appeared (B. L. Browning, ed., "The Chemistry of Wood," Wiley-Interscience, New York, 1963). Two years later, Sven Rydholm in his book "Pulping Processes" (Wiley-Interscience, New York, 1965) gave a unique exposé of pulping, discussing this matter comprehensively also from the standpoint of wood chemistry. The present book has been written in the belief that there is now a rather wide circle of readers who need a knowledge of modern wood chemistry in the form of a textbook. In addition to pulping and papermaking there are numerous potential applications in wood chemistry particularly connected with the utilization of wood and wood wastes as well as the by-products from pulping processes for production of chemicals and energy. Indeed, as a renewable raw material, wood constitutes an enormous resource for biomass conversion in the future.

This book attempts to discuss various aspects of wood chemistry in relation to applications. It is believed that the book might be useful not only for students and teachers but generally for chemists, biochemists, and others working either in the laboratory as researchers or in production and planning. Chapter 1 describes the structure and anatomy of wood. Carbohydrate chemistry belongs to the fundamentals in wood chemistry because two thirds of the wood constituents are polysaccharides. Chapter 2 therefore deals with the general structure, properties, and pertinent reactions of carbohydrates. In Chapter 3 the chemistry and polymer properties of wood polysaccharides are specifically discussed. The challenging chemistry of lignin is presented in Chapter 4 in conjunction with morphological aspects. Chapter 5 covers the interesting group of extractives, which consists of extremely diversified constituents. They cause problems in pulping and bleaching, but are also a source of valuable by-products. The anatomy and chemistry of bark also are discussed (Chapter 6). The reactions of wood constituents during sulfite and kraft pulping and bleaching are dealt with in Chapters 7 and 8. Basic inorganic reactions of the pulping and bleaching chemicals are included in addition to some general aspects pertinent to the technology of the delignification processes. Chapter 9 covers cellulose derivatives and related products. Chapter 10 finally discusses various alternatives and possibilities for utilization of solid wood (residues) as well as by-products from pulping. It is hoped that this challenging field will attract chemists for new endeavors.

Based on current concepts an attempt has been made to present a rationalized and logical account of wood chemistry with emphasis on its applications. Although not covered comprehensively, references to the relevant literature are listed at the end of each chapter. Many of these are only examples selected from the vast collection available and serve those readers who need a further guidance or information of the topics discussed.

Although much of the content is based on my earlier book in Finnish on wood chemistry, the first edition of which appeared in 1977 (Otakustantamo, Espoo), this version has been considerably improved and enlarged. Fortunately, at the early stages of the preparation of the manuscript, I was encouraged by Professors K. V. Sarkanen and T. E. Timell, who generously offered their help. I am deeply indebted to them for checking the manuscript in detail and for the numerous improvements with respect to both content and language. Other friends and colleagues, including Dr. W. Brown, Professors J. Gierer, J. Gripenberg, J. J. Lindberg, T. Norin, B. Rånby, O. Theander, and Drs. J. Janson, K. Kringstad, and B. Lindgren, read portions of the manuscript and offered many useful comments. I am also grateful for the material provided by Dr. E. Back, Professor W. A. Côté, Professor D. P. Delmer, Dr. D. A. I. Goring, Mrs. M-S. Ilvessalo-Pfäffli, Professor H. Meier, and

Professor T. E. Timell. Finally, I wish to thank the staff members of our laboratory, R. Alén, Christine Hagström, E. Seppälä, and T. Vuorinen who provided help in several respects and Kristiina Holm for typing and Eija Wiik and Ritva Valta for the drawings.

Eero Sjöström

Chapter 1

THE STRUCTURE OF WOOD

Trees belong to seed-bearing plants (Spermatophytae), which are sub-divided into gymnosperms (Gymnospermae) and angiosperms (Angiosper-mae). Coniferous woods or softwoods belong to the first-mentioned category and hardwoods to the second group. Altogether 30,000 angiosperms and 520 coniferous tree species are known; most of the former grow in tropical forests. In North America the number of species is about 1200, while in Europe only 10 softwood and 51 hardwood species exist naturally. This limited number represents species surviving the period of glaciation, during which genera such as *Sequoia* and *Pseudotsuga* completely disappeared from Europe.

1.1   The Macroscopic Structure of Wood

Wood is composed of elongated cells, most of which are oriented in the longitudinal direction of the stem. They are connected with each other through openings, referred to as pits. These cells, varying in their shape according to their functions, provide the necessary mechanical strength to the tree and also perform the function of liquid transport as well as the storage of reserve food supplies.

Figure 1-1 shows the macroscopic structure of wood as it appears to the

1

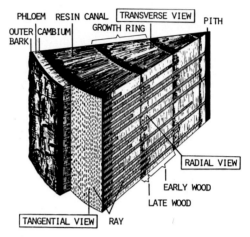

Fig. 1-1.   Sections of a four-year-old pine stem.

naked eye. The centrally located pith is descernible as a dark stripe in the middle of the stem or branches. It represents the tissues formed during the first year of growth. The xylem or wood is organized in concentric growth rings (annual increments). It also contains rays in horizontal files, extending from the outer bark either to the pith (primary rays) or to an annual ring (secondary rays). Some softwoods also contain resin canals. The inner part of a tree usually consists of dark-colored heartwood. The outer part, or sapwood, is lighter in color and conducts water from the roots to the foliage of the tree. The cambial zone is a very thin layer consisting of living cells between the wood (xylem) and the inner bark (phloem). The cell division and radial growth of the tree takes place in this region.

1.2   The Living Tree

1.2.1   The Growth of the Tree

The tree grows through the division of the cells. The length of the growth period largely depends on the climate, but in many parts of North America and Scandinavia growth occurs from May to early September, and is most intensive in the spring. The majority of the cells develop into various permanent cells and only a very few are retained as growing cells capable of division.

The growth of a tree is always continuous although it becomes slower in the course of time. Giant sequoias (*Sequoiadendron giganteum*) in California can be up to 4000 years old measuring 100 meters in height and 12 meters in diameter at the base.

Longitudinal growth (primary growth), which takes place in the early season, proceeds at the end of the stem, branches and roots. The growth points are located inside the buds, which have been formed during the preceding autumn.

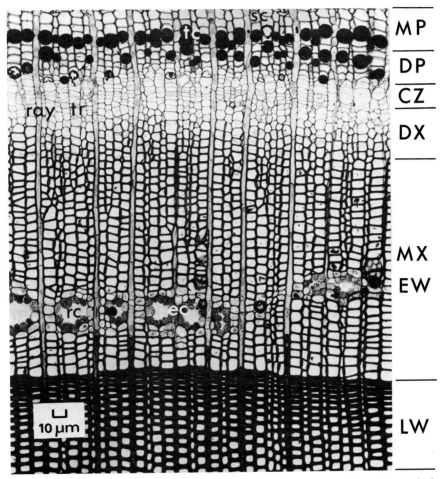

Fig. 1-2. Transverse section of xylem and phloem of red spruce (*Picea rubens*). CZ, cambial zone; DP, differentiating phloem; MP, mature phloem with sieve cells (sc) and tannin cells (tc); DX, differentiating xylem with ray cells and tracheids (tr); MX, mature xylem, earlywood (EW) with resin canals (rc), lined with epithelial cells (ec); LW, latewood. Note that each ray continuous from the xylem, through the cambial zone, and into the phloem. Light micrograph by L. W. Rees. Courtesy of Dr. T. E. Timell.

Radial growth begins in the *cambium* which is composed of a single layer of thin-walled living cells (initials) filled with protoplasm (cf. Fig. 1-2). The *cambial zone* consists of several rows of cells, which all possess the ability to divide. On division the initial cell produces a new initial and a xylem mother cell, which in its turn gives rise to two daughter cells; each of the latter is capable of further division. More cells are produced toward the xylem on the inside than toward the phloem on the outside; phloem cells divide less frequently than xylem cells. For these reasons, trees always contain much more wood than bark.

1.2.2   Development of the Cell

When a cell divides, it first develops a cell plate, which is rich in pectic substances. Each of the two new cells subsequently encloses itself with a thin, extensible, primary wall, consisting of cellulose, hemicelluloses, pectin, and protein. During the following phase of differentiation, the cell first expands to its full final size, after which formation of the thick, secondary wall is initiated. At this stage, this wall consists of cellulose and hemicelluloses. Lignification begins while the secondary wall is still being formed. Figs. 1-3 and 1-12 show the structures of a mature cell.

1.2.3   Annual Rings

At the beginning of the growth the tree requires an effective water transportation system. In softwoods thin-walled cells with large cavities are formed; in hardwoods special vessels take care of the liquid transportation. Comparatively light-colored and porous *earlywood* is thus formed. Later, the rate of growth decreases and *latewood* is produced. It consists of thick-

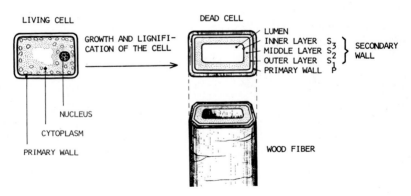

Fig. 1-3. Development of the living cell to wood fiber (Bucher, 1965).

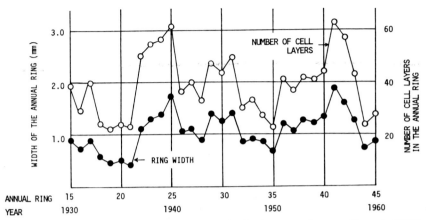

Fig. 1-4. Annual variations in the radial growth of a slowly grown pine stem (Ilvessalo-Pfäffli, 1967).

walled fibers and gives mechanical strength to the stem and is darker and denser than the earlywood.

The age of a tree can be calculated from the number of growth rings at the base of the stem. With a continuous growth period (tropical woods) regular annual rings are lacking. Alternation of wet and dry periods may, however, result in the formation of growth rings. The boundary between earlywood and latewood varies. It may be very sharp as in larch or nearly nonexisting (birch, aspen, and alder). Earlywood is weaker than the thick-walled latewood. Pulp fibers from earlywood and latewood also have different papermaking properties.

The width of the annual rings varies greatly depending on tree species and growth conditions. The variation limits for Scots pine in Scandinavia may be 0.1–10 mm (Fig. 1-4). For similar reasons the proportion of latewood may vary greatly. Typical percentages for the latewood in Scandinavia are 15–50% for pine and 10–40% for spruce; the values are higher in the northern than in the southern parts of these countries.

1.2.4 Cell Types

On the basis of their different shape wood cells can be divided into prosenchyma and parenchyma cells. The former are thin, long cells, narrower toward the ends; the latter are rectangular or round and are short cells.

Depending on their functions, cells can be divided into three different groups: conducting cells, supporting cells, and storage cells. Conducting and supporting cells are dead cells containing cavities which are filled with water or air. In hardwoods the conducting cells consist of vessels and the supporting cells of fibers. In softwoods the tracheids perform both functions. The storage cells transport and store nutrients. They are thin-walled parenchyma cells which function as long as they remain in the sapwood.

1.2.5 Pits

Water conduction in a tree is made possible by pits, which are recesses in the secondary wall between adjacent cells. Two complementary pits normally occur in neighboring cells thus forming a pit pair (Fig. 1-5). Water transport between adjacent cell lumina occurs through a pit membrane which consists of a primary wall and the middle lamella. Bordered pit pairs are typical of softwood tracheids and hardwood fibers and vessels. In softwoods the pit membrane might be pressed against the pit border thus preventing water transport, since the torus is impermeable. The pits connecting tracheids, fibers, and vessels with the ray parenchyma cells are half-bordered. Simple pits without any border connect the parenchyma cells with one another.

The different shape of the pits are distinctive features in the microscopic identification of wood and fibers. Knowledge of the porous structure of wood is also of great importance for understanding the phenomena which are associated with the impregnation of wood.

1.2.6 Softwood cells

The wood substance in softwoods is composed of two different cells: tracheids (90–95%) and ray cells (5–10%).

Tracheids give softwoods the mechanical strength required (especially the thick-walled latewood tracheids) and provide for water transport, which

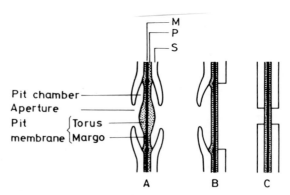

Fig. 1-5. Types of pit pairs. A, bordered pit pair; B, half bordered pit pair; C, simple pit pair; M, middle lamella; P, primary wall; S, secondary wall.

occurs through the thin-walled early wood tracheids with their large cavities. The liquid transport from one tracheid to another takes place through the bordered pits; their amount in earlywood tracheids is about 200 per tracheid, most of them located in the radial walls in one to four lines. Latewood tracheids have only 10 to 50 rather small bordered pits.

Liquids move from the tracheids to the ray parenchyma cells through half-bordered pits. The location and nature of these pits are characteristic and used for the identification of different wood species (compare the small

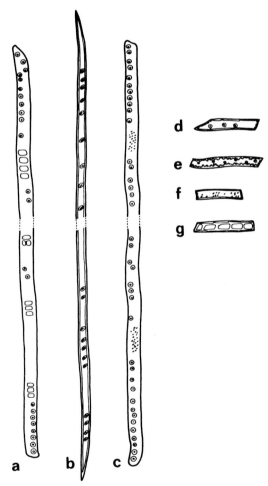

Fig. 1-6. Cells of coniferous woods. An earlywood (a) and a latewood (b) pine tracheid, an earlywood spruce tracheid (c), ray tracheid of spruce (d) and of pine (e), ray parenchyma cell of spruce (f) and pine (g) (Ilvessalo-Pfäffli, 1967).

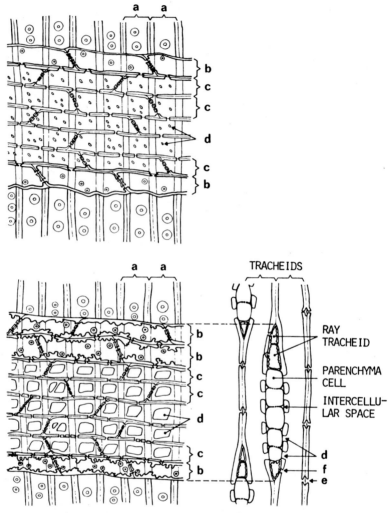

Fig. 1-7.   Radial section of a spruce ray (above) and radial and tangential section of a pine ray (below). (a) Longitudial tracheids. (b) Rows of ray tracheids (small bordered pits). (c) Rows of ray parenchyma. (d) Pits in the cross fields leading from ray parenchyma to longitudial tracheids. (e) A bordered pit pair between two tracheids. (f) A bordered pit pair between a longitudial and a ray tracheid (Ilvessalo-Pfäffli, 1967).

elliptic pits in spruce with the large window pores in Scots pine, Figs. 1-6 and 1-7).

As in other cells, the dimensions of the tracheids vary depending on genetic factors and growth conditions. Variations exist among different species and individuals as well as between different parts of the stem and within one and the same growth ring. The fiber length in the stem increases from the pith toward the cambium and reaches a maximum at the middle of the bole. Tracheids in the latewood or in narrow annual rings are usually longer and narrower than those formed more rapidly. The tangential width of the fibers varies only slightly but large differences exist in the radial direction between earlywood and latewood tracheids.

The average length of Scandinavian softwood tracheids (Norway spruce and Scots pine) is 2-4 mm and the width in the tangential direction is 0.02-0.04 mm (Fig 1-8). The thickness of earlywood and latewood tracheids is 2-4 μm and 4-8 μm, respectively.   ✴ wall

The width of a ray usually corresponds to one cell. Several parenchyma cell files are placed on top of one another. Ray tracheids are often located at the upper and lower edges of this tier (Fig. 1-7). Parenchyma cells are thin-walled, living cells. In Norway spruce and Scots pine their length and width vary between 0.01-0.16 mm and 2-50 μm, respectively. The ray tracheids are of the same size and also provide liquid transport in the radial direction. Rays in Scots pine, for example, contain 25-31 ray tracheids per square millimeter in a tangential section.

Resin canals are intercellular spaces building up a uniform channel network in the tree. Horizontal canals are always located inside the rays which

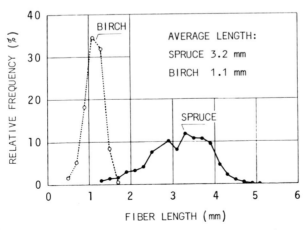

Fig. 1-8. Example of the distribution of fiber length in softwood (*Picea abies*) and hardwood (*Betula verrucosa*) (Ilvessalo-Pfäffli, 1977).

appear together in several files (fusiform rays) (Fig. 5-2). The resin canals are lined by epithelial parenchyma cells, which secrete oleoresin into the canals. Pine wood contains more and larger resin canals than does spruce wood. In pine they are concentrated in the heartwood and root, whereas in spruce they are evenly distributed throughout the whole wood. The diameters of the resin canals in pine are on the average about 0.08 mm (vertical) and 0.03 mm (radial) (see Section 5.1.1).

1.2.7   Hardwood Cells

Hardwoods contain several cell types, specialized for different functions (Fig. 1-9). The supporting tissue consists mainly of libriform cells, the conducting tissue of vessels with large cavities, and the storage tissue of ray parenchyma cells. In addition, hardwood contains hybrids of the above-mentioned cells which are classified as fiber tracheids. Although the term *fiber* is frequently used for any kind of wood cells, it more specifically denotes the supporting tissue, including both libriform cells and fiber tracheids. In birch these cells constitute 65 to 70% of the stem volume.

Libriform cells are elongated, thick-walled cells with small cavities containing some simple pits. The dimensions of birch libriform fibers are 0.8–1.6 mm or on an average 1.1–1.2 mm (length), 14–40 μm (width), and 3–4 μm

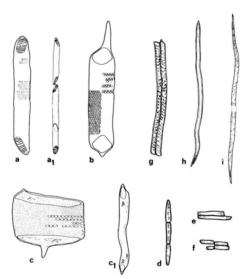

Fig. 1-9.   Hardwood cells. Vessel elements of birch (a), of aspen (b), and of oak in earlywood (c) and in latewood (c_1), as well as a birch vessel (a_1). Longitudinal parenchyma of oak (d) and ray parenchyma of aspen (e) and of birch (f). Tracheids of oak (g) and birch (h) and a birch libriform fiber (i) (Ilvessalo-Pfäffli, 1967).

(cell wall thickness). In some tropical hardwood species the average length may reach 4 mm.

Vessels are composed of thin-walled and rather short (0.3–0.6 mm) and wide (30–130 μm) elements, which are placed on top of one another to form a long tube. The ends have disappeared more or less completely. The channels thus formed, which might be several meters in length, are capable of a more effective water transport than the softwood tracheids. This is needed especially in the spring during the leafing. In diffuse-porous woods (aspen, birch, and maple) the vessels are evenly distributed across the annual ring. The vessels are larger and more numerous in the earlywood portion in ring-porous woods, such as ash, elm, and oak. In birch and aspen the vessels amount to about 25% of the wood volume. Several different pores are present in the walls of the vessels. These differences together with other structural features are of great help in the identification of pulp fibers. Besides the usual vessels some hardwoods contain cells resembling softwood tracheids or small vessels. Their walls are rich in bordered pits.

Hardwood rays consist exclusively of *parenchyma cells*. The ray width varies in the tangential direction. In aspen wood the rays form one row, in birch wood and oak wood 1–3 and 1–30 rows, respectively. The height varies from one up to several hundred tiers. The rays account for 5–30% of the stem volume.

1.2.8 Sapwood and Heartwood

At a certain age the inner wood of the stem of most trees begin to change to a completely dead heartwood and its proportion of the stem becomes successively larger as the tree grows. The dying parenchyma cells produce organic deposits such as resin, phenolic substances, pigments, etc. In softwoods the bordered pits are closed when the torus becomes pressed against either side of the border. In some hardwoods, such as oak or ash, the vessels are closed by tyloses, which enter the vessel from neighboring ray cells (Fig. 1-10). Wood with tyloses is impermeable to liquids and an excel-

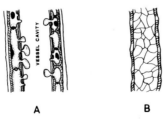

A   B

Fig. 1-10. Tyloses at the bud stage (A) and at a later stage filling the vessel cavity (B), both in lateral view.

lent material for barrels. These anatomical and chemical changes often have a significant influence on the behavior of sapwood and heartwood during pulping.

1.3   Wood Ultrastructure

1.3.1   Building Elements

The wood cell consists mainly of cellulose, hemicelluloses, and lignin (see Appendix). A simplified picture is that cellulose forms a skeleton which is surrounded by other substances functioning as matrix (hemicelluloses) and encrusting (lignin) materials.

The length of a native cellulose molecule is at least 5000 nm corresponding to a chain with about 10,000 glucose units (cf. Section 3.2.2). The smallest building element of the cellulose skeleton is considered by some to be an elementary fibril. This is a bundle of 36 parallel cellulose molecules which are held together by hydrogen bonds, but various opinions exist concerning this question. The cellulose molecules according to the "fringe micellar model" form completely ordered or crystalline regions, which without any distinctive boundary are changing into disordered or amorphous regions (Fig. 1-11). In native cellulose the length of the crystallites can be 100–250 nm and the cross section, probably rectangular, is on an average 3×10 nm. According to this model the cellulose molecule continues through several crystallites.

The microfibrils, which are 10–20 nm wide, are visible in the electron microscope without pretreatment. Microfibrils are combined to greater fibrils and lamellae, which can be separated from the fibers mechanically, although their dimensions greatly depend on the method used.

Disordered cellulose molecules as well as hemicelluloses and lignin are located in the spaces between the microfibrils. The hemicelluloses are considered to be amorphous although they apparently are oriented in the same direction as the cellulose microfibrils. Lignin is both amorphous and isotropic.

Fig. 1-11.   Diagrammatic representation of fibrillar structure in the cell wall according to Mark (1940). Heavy lines constitute the crystalline regions. The chain molecules may pass through one or more crystalline and amorphous regions.

1.3.2 Cell Wall Layers

The cell wall is built up by several layers, namely (Figs. 1-12 and 1-13), middle lamella (M), primary wall (P), outer layer of the secondary wall (S_1), middle layer of the secondary wall (S_2), inner layer of the secondary wall (S_3), and warty layer (W). These layers differ from one another with respect to their structure as well as their chemical composition. The microfibrils wind around the cell axis in different directions either to the right (Z helix) or to the left (S helix). Deviations in the angular directions cause physical differences and the layers can be observed in a microscope under polarized light.

The *middle lamella* is located between the cells and serves the function of binding the cells together. At an early stage of the growth it is mainly composed of pectic substances, but it eventually becomes highly lignified. Its thickness, except at the cell corners, is 0.2–1.0 μm. The *primary wall* is a thin layer, 0.1–0.2 μm thick, consisting of cellulose, hemicelluloses, pectin,

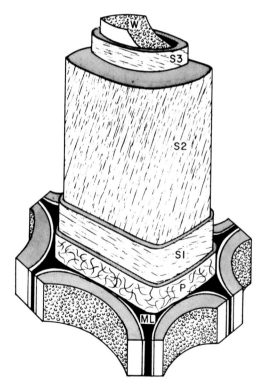

Fig. 1-12. Simplified structure of a woody cell, showing the middle lamella (ML), the primary wall (P), the outer (S_1), middle (S_2), and inner (S_3) layers of the secondary wall, and the warty layer (W) (Côté, 1967, with permission).

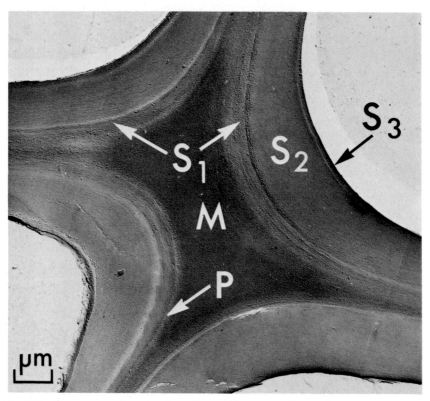

Fig. 1-13. Transverse section of earlywood tracheids in tamarack (*Larix laricina*), showing the middle lamella (M), the primary wall (P), and the outer (S₁), middle (S₂), and inner (S₃) layers of the secondary wall. Transmission electron micrograph. Courtesy of Dr. T. E. Timell.

and protein and completely embedded in lignin. The cellulose microfibrils form an irregular network in the outer portion of the primary wall; in the interior they are oriented nearly perpendicularly to the cell axis (Fig. 1-14). In the presence of reagents which induce strong swelling the primary wall is peeled off and the belts around the fibers expand (*ballooning*) (Fig. 1-15). The middle lamella together with the primary walls on both sides, is often referred to as the compound middle lamella. Its lignin content is high, but because the layer is thin only 20–25% of the total lignin in wood is located in this layer.

The *secondary wall* consists of three layers: thin outer and inner layers and a thick middle layer. These layers are built up by lamellae formed by almost parallel microfibrils between which lignin and hemicelluloses are located.

The *outer layer* (S_1) is 0.2–0.3 μm thick and contains 3–4 lamellae where

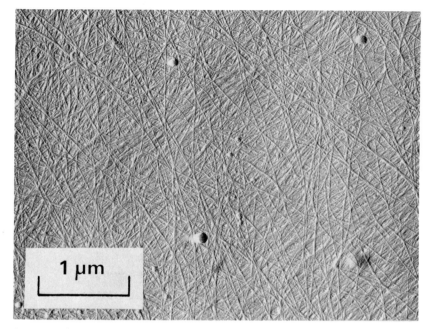

Fig. 1-14. Electron micrograph of a delignified primary wall (*Pinus sylvestris*) (Meier, 1958).

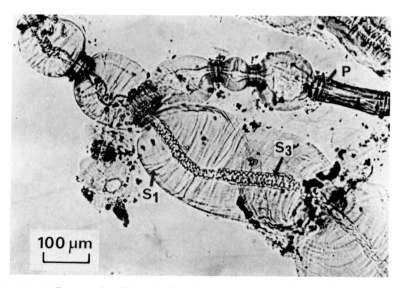

Fig. 1-15. Ballooning of a sulfate pulp fiber (*Pinus sylvestris*). The ribbonlike, unrolled primary wall (P) surrounds the swollen secondary wall. S_1 is the swollen outer layer of the secondary wall, under which the microfibrils of the middle layer, nearly parallel to fiber axis, are dimly visible. S_3 is the inner layer of the secondary wall (Ilvessalo-Pfäffli, 1977).

the microfibrils form either a Z helix or S helix. The microfibril angle of the crossed fibrillar network varies between 50 and 70° with respect to the fiber axis.

The *middle layer* (S_2) forms the main portion of the cell wall. Its thickness in softwood tracheids varies between 1 (earlywood) and 5 (latewood) μm and it may thus contain 30–40 lamellae or more than 150 lamellae. The thickness naturally varies with the cell types. The microfibrillar angle (Fig. 1-16) varies between 10° (earlywood) and 20–30° (latewood). It decreases in a regular fashion with increasing fiber length. The characteristics of the S_2 layer (thickness, microfibrillar angle, etc.) have a decisive influence on the fiber stiffness as well as on other papermaking properties.

The *inner layer* (S_3) is a thin layer (ca. 0.1 μm) consisting of several lamellae which contain microfibrils in both Z helices and S helices (50–90° angle). Great variations are noted among different wood species.

The *warty layer* (W) is a thin amorphous membrane located in the inner surface of the cell wall in all conifers and in some hardwoods, containing warty deposits of a still unknown composition. Each species has its own, characteristic warty layer.

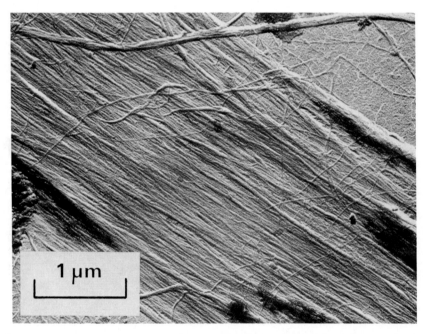

Fig. 1-16.   Electron micrograph of a delignified secondary wall (S_2) of *Pinus sylvestris.* Courtesty of Dr. H. Meier.

1.3.3 Pits

The normal structure of the cell wall is broken by pits. Changes appear already in the growth period of the cell. For instance, early stages of pit formation in softwoods are visible in the primary wall just before the cell reaches its final dimensions (primary pit fields). The microfibril network is loosened and new microfibrils are oriented around these points. The structure in the middle of the circles is tightened and the radially oriented microfibril bundles finally form a netlike membrane, permeable to liquids (*margo*) (Fig. 1-17). The central, thickened portion of the pit membrane

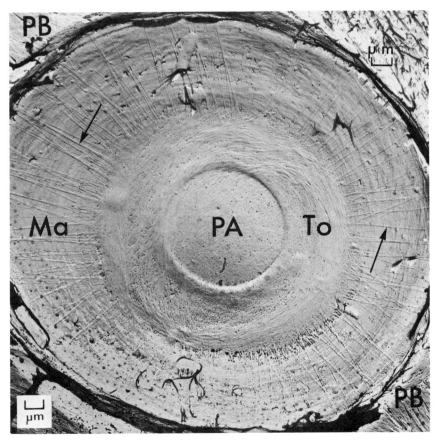

Fig. 1-17. Surface replica of an aspirated bordered pit in a tracheid of Douglas fir (*Pseudotsuga menziesii*), showing the pit aperture (PA), the torus (To), the margo (Ma), and the pit border (PB). Arrows indicate supporting cellulose strands. Transmission electron micrograph. Courtesy of Dr. W. A. Côté, Jr.

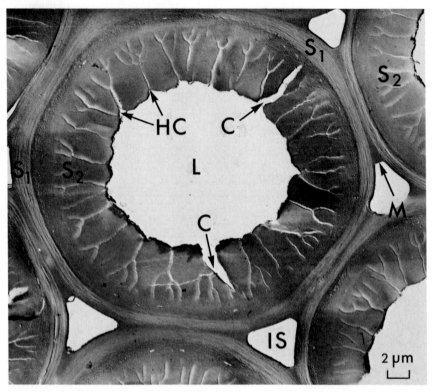

Fig. 1-18. Transverse section of compression wood tracheids in tamarack (*Larix laricina*), showing intercellular spaces (IS), middle lamella (M), the outer (S₁), and the inner (S₂) layer of the secondary wall, and the lumen (L). The S₂ layer contains narrow, branched helical cavities (HC) as well as two wide drying checks (C), an artifact. Transmission electron micrograph. Courtesy of Dr. T. E. Timell.

(*torus*) is formed after a secondary thickening of the primary wall. The torus is rich in pectic material and also contains cellulose in pine and spruce.

1.4   Reaction Wood

As a product of living organism the structure of wood fibers is so variable and complicated that a great number of details remains to be solved for understanding the anatomy and biology even of trees grown under normal conditions. When a tree is brought out of its natural, equilibrium position in space, for example by wind or by a landslide, the tree begins to produce a special tissue, referred to as *reaction wood*. The function of this type of wood is to restore the displaced stem or branch to its original position. In a leaning stem of a conifer, *compression wood* develops on the lower side. This wood expands longitudinally as it is being formed, and the pressure exerted along the grain forces the stem to bend upward. All movements of orientation in mature conifers are effected with the aid of appropriately located compression wood. In hardwoods, *tension wood* is formed on the upper side of an inclined stem. This wood contracts as it is laid down and in this way forces the stem to bend upward. Compression wood can be said to *push* a stem or a branch up; tension wood *pulls* them up.

Compression wood is heavier, harder, and denser than the normal wood. Its tracheids are short and thick-walled (even in earlywood) and in cross section rounded so that empty spaces remain between the cells. The S_1 layer is thicker than in a normal wood while the S_3 layer is absent. The S_2 layer contains helical cavities that parallel the microfibrils and reach from the

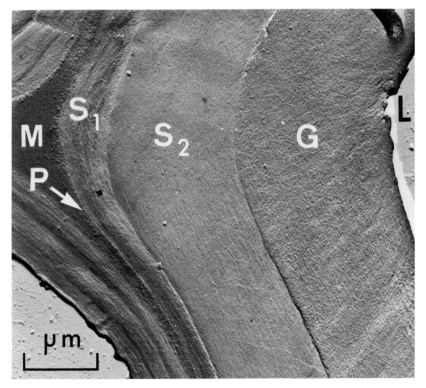

Fig. 1-19. Transverse section of a tension wood fiber in American beech (*Fagus grandifolia*), showing the middle lamella (M), primary wall (P), the outer (S_1) and middle (S_2) layers of the secondary wall, the thick gelatinous layer (G), and the lumen (L). Transmission electron micrograph. Courtesy of Dr. T. E. Timell.

lumen deep into the S_2 (Fig. 1-18). The cellulose content of compression wood is lower and the lignin content higher than for normal wood.

Tension wood differs less from normal wood than compression wood. It contains thick-walled fibers, terminated towards the lumen by a gelatinous layer (Fig. 1-19). This so-called G layer consists of pure and highly crystalline cellulose oriented in the same direction as the fiber axis. For this reason the cellulose content of tension wood is higher and the lignin content lower than in normal wood.

References

Bucher, H. (1965). Das Geheimnis des Holzes. *Hespa Mitt.* **15**(3), 1–24.

Core, H. A., Côté, W. A., and Day, A. C. (1979). "Wood: Structure and Identification," 2nd ed. Syracuse Univ. Press, Syracuse, New York.

Côté, W. A., Jr., ed. (1965). "Cellular Ultrastructure of Woody Plants." Syracuse Univ. Press, Syracuse, New York.

Côté, W. A., Jr. (1967). "Wood Ultrastructure." Univ. of Washington Press, Seattle.

Frey-Wyssling, A., and Mühlethaler, K. (1965). "Ultrastructural Plant Cytology (with an Introduction to Molecular Biology)." Elsevier, Amsterdam.

Ilvessalo-Pfäffli, M.-S. (1967). The structure of wood. *In* "Wood Chemistry" (W. Jensen, ed.), 1st ed., Vol. 1(B 1), pp. 1–50. Text Handb. Finn. Pap. Eng. Assoc., Helsinki. (In Finn.)

Ilvessalo-Pfäffli, M.-S. (1977). The structure of wood. *In* "Wood Chemistry" (W. Jensen, ed.), 2nd ed., Vol. 1, pp. 7–81. Text Handb. Finn. Pap. Eng. Assoc., Helsinki. (In Finn.)

Mark, H. (1940). Intermicellar hole and tube system in fiber structure. *J. Phys. Chem.* **44**, 764–787.

Meier, H. (1958). The fine structure of wood fibers. *Sven. Papperstidn.* **61**, 633–640. (In Swed.)

Panshin, A. J., and de Zeeuw, C. (1980). "Textbook of Wood Technology, Vol. 1, Structure, Identification, Uses and Properties of the Commercial Woods of the United States and Canada," 4th ed. McGraw-Hill, New York.

Timell, T. E. (1973). Ultrastructure of the dormant and active cambial zones and the dormant phloem associated with formation of normal and compression woods in *Picea abies* (Karst.). *Tech. Publ.—State Univ. N.Y., Coll. Environ. Sci. For., Syracuse* No. 96, pp. 1–23.

Tsoumis, G. (1968). "Wood as Raw Material." Pergamon, Oxford.

Chapter 2

FUNDAMENTALS IN CARBOHYDRATE CHEMISTRY

2.1 The Concept and Nomenclature of Carbohydrates

The name carbohydrate was originally derived from the general formula $C_x(H_2O)_y$, formally analogous to hydrates of carbon, but this analogy is rather misleading. The carbohydrates are actually polyhydroxy compounds appearing commonly in nature, either as relative small molecules (sugars) or as large entities extending to macromolecular levels (polysaccharides). Sugars are formed in green plants as early products of photosynthesis from carbon dioxide and water and are then converted into organic plant constituents through a variety of biosynthetic pathways.

The sugars in a plant usually function as a source of energy while polysaccharides, such as starch, fulfill the need for the storage of reserve food or they (cellulose and hemicelluloses) contribute mechanical strength to the plant cell wall. In addition, a variety of carbohydrates are included as essential building elements in natural compounds performing vital functions in living organisms.

The carbohydrates may be calssified into following three large groups: (1) *Monosaccharides* are simple sugars, of which glucose, mannose, galactose, xylose, and arabinose are the most common constituents of the cell wall polysaccharides in wood. (2) *Oligosaccharides* consist of several

TABLE 2-1. Aldose Series[a]

Carbon atom number	Trioses	Tetroses	Pentoses	Hexoses
1	CHO	CHO	CHO	CHO
2	*CHOH	*CHOH	*CHOH	*CHOH
3	CH$_2$OH	*CHOH	*CHOH	*CHOH
4		CH$_2$OH	*CHOH	*CHOH
5			CH$_2$OH	*CHOH
6				CH$_2$OH

[a] Asterisk denotes asymmetric carbon atom.

monosaccharide residues joined together by glycosidic linkages, named di-, tri-, tetrasaccharides, and so on. The name oligosaccharide is usually restricted to the group of carbohydrates in which the number of monosaccharide units is less than ten. (3) *Polysaccharides* are complex molecules composed of a large number of monosaccharide units joined together by glycosidic linkages. *Polyuronides* are polysaccharides containing uronic acid blocks in the main backbone and are typical components in algae (sea weeds) and pectins. *Polyhydric alcohols,* consisting of acyclic polyols (alditols, glycitols, or "sugar alcohols") and alicyclic polyalcohols (cyclitols), are classified as carbohydrates in a wider sense. Of the former sorbitol (D-glucitol) occurs in algae as well as in higher plants and was discovered in the fresh juice of berries of mountain ash (*Sorbus aucuparia*). Of cyclitols cyclohexanehexols or inositols, particularly *myo*-inositol, have a wide distribution and are of importance for plants, bacteria, and animals.

The monosaccharides contain either an aldehyde or a keto function and are accordingly classified as *aldoses* or *ketoses*. The aldoses and ketoses are further divided into subgroups on the basis of their number of carbon atoms. The major carbohydrates in wood consist of aldopentoses and aldohexoses (Table 2-1). A prefix deoxy is used when one of the hydroxyl groups is replaced by a hydrogen atom ($\equiv$C—OH → $\equiv$C—H). An ether group, usually methyl ($\equiv$C—OH → $\equiv$C—OMe), is denoted by a prefix (*O*-methyl). Correspondingly, the prefix for an ester group, such as acetate ($\equiv$C—OH → $\equiv$C—OCOCH$_3$) is *O*-acetyl.

2.2 Monosaccharides

Most of the monosaccharides occur as glycosides and as units in oligosaccharides and polysaccharides and only comparatively few of them are present free in plants. Glucose is the most abundant monosaccharide in nature. It occurs in a free state in many plants, especially in fruits and can be prepared from cellulose and starch by acidic or enzymic hydrolysis. Of the other aldohexoses mannose and galactose are important components in hemicelluloses. The most common representatives of aldopentoses are xylose and arabinose. Ribose is a constituent of nucleosides. No tetroses or trioses have been detected free in plants, but D-erythrose-4-phosphate is an important intermediate in many transformations and D-glyceraldehyde and dihydroxyacetone are essential components in cellular metabolism. Of the heptoses, sedoheptulose-7-phosphate occurs as intermediate in photosynthesis and traces of it may be present in all plants. Of the deoxysugars, rhamnose (6-deoxymannose), occurs as constituent in gum polysaccharides and traces of it are present in hardwood hemicelluloses. Fructose, which represents the only abundant ketose in plants, is present both free and in a combined state. Plants belonging to the Compositeae and Gramineae families store polymers of fructose, such as inulin, as reserve material rather

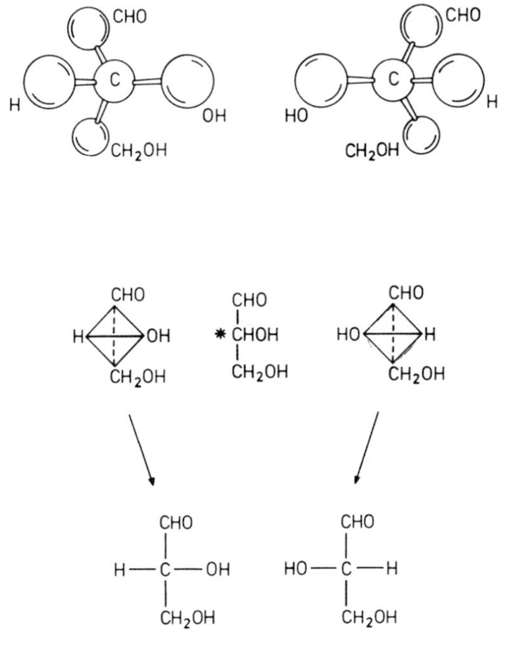

Fig. 2-1. Representation of the D and L forms of glyceraldehyde (1 and 2). Note that in the Fischer projection formulas the substituents oriented upward from the plane (H and OH) are horizontal whereas the substituents oriented downward (CHO and CH$_2$OH) are vertical.

than starch. Fructose is obviously not present in the cell wall polysaccharides of wood.

2.2.1 The Configuration of Monosaccharides

An example of a simple sugar, aldotriose or glyceraldehyde is given in Fig. 2-1. It contains one "*asymmetric carbon atom*" or so-called *chiral center* since all the four substituents bound to this carbon atom are different. It follows from this that glyceraldehyde can exist in two stereoisomeric forms, which are mirror images of each other and termed *enantiomers*. These enantiomers cause rotation of the plane of plane-polarized light to an equal degree but in opposite directions. The enantiomeric forms of glyceraldehyde can be visualized by a tetrahedron in which the substituents appear at the vertices, as originally suggested by van't Hoff and later adopted in carbohydrate chemistry by Emil Fischer. A simplification of this concept is the *Fischer projection formula,* which is a two-dimensional representation of the molecule. In this the bond between the carbon atoms is vertical and the bonds connecting H and OH groups to the carbon are horizontal. The symbol D is assigned to (+)-glyceraldehyde since it has the hydroxyl group on the right in the Fischer projection formula whereas L refers to (−)-glyceraldehyde which has the hydroxyl group on the left. Figure 2-2 shows the Fischer projection formula for D-xylose.

Any molecule with n asymmetric carbon atoms can exist in the form of 2^n stereoisomers including 2^{n-1} enantiomeric (mirror image) pairs. When the number of asymmetric carbon atoms exceeds two, so-called *diastereoisomeric* forms become possible. They possess different physical properties and are not mirror images. *Enantiomers,* however, are identical in physical properties with the exception of their behavior toward polarized light. The aldohexoses comprise 16 stereoisomers (8 enantiomeric pairs) belonging to the respective series. Within the D or L series the individual aldohexoses are diastereoisomers. The aldoses in D series are shown in Fig. 2-3.

In some cases a compound is optically inactive although it contains chiral

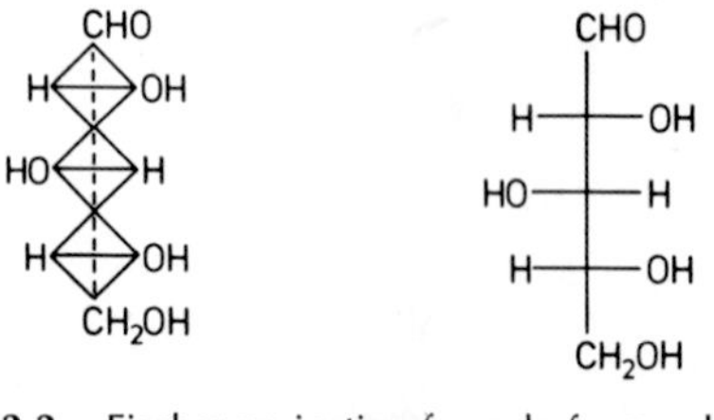

Fig. 2-2. Fischer projection formula for D-xylose.

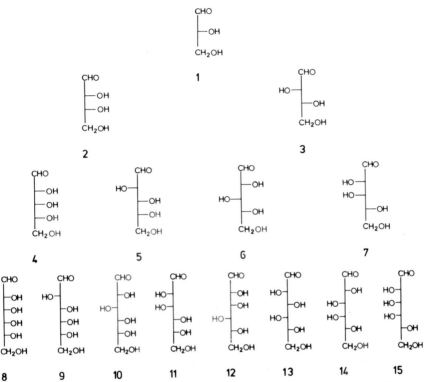

Fig. 2-3. Acyclic forms of the D-series of aldoses. 1, Glyceraldehyde; 2, erythrose; 3, threose; 4, ribose; 5, arabinose; 6, xylose; 7, lyxose; 8, allose; 9, altrose; 10, glucose; 11, mannose; 12, gulose; 13, idose; 14, galactose; 15, talose.

centers since it is superimposable upon its mirror image. Such an optically inactive stereoisomer is designated by the prefix *meso*.

According to the Rosanoff's convention the configuration is determined by the orientation of the hydroxyl group bound to the last asymmetric carbon atom in the carbon chain, which, for example in glucose is C-5. In the D form this hydroxyl group points to the right in the Fischer projection formula whereas in the L form the hydroxyl group is on the left.* The symbols D and L specify the *absolute configuration* and they bear no relationship to the direction of optical rotation, which can be separately marked by (+) or (−) after the configuration symbols.

*The use of DL-convention and the Fischer projection for absolute configuration has some disadvantages. According to a later and a more exact convention (Cahn-Ingold-Prelog) the configuration of a chiral center is specified using prefixes *R* and *S* (Latin: *rectus*, right; *sinister*, left). In carbohydrate chemistry, however, the DL-convention is used generally.

Fig. 2-4. Formation of a cyclic hemiacetal. Rings are usually five- or six-membered.

2.2.2 The Ring Structures of Monosaccharides

Aldehydes and ketones are able to form hemiacetals with alcohols. Since sugar molecules contain an aldehyde or a keto group as well as hydroxyl groups, a cyclic *hemiacetal* is readily formed, which in solutions is in equilibrium with the open-chain form (Fig. 2-4). Hemiacetal rings are composed of five or six atoms; smaller or larger rings are too strained or not stable for thermodynamic reasons with the exception of a seven-membered ring (*septanose*), which occasionally can occur. For example, glucose and fructose can exist both as a six-membered (*pyranose*) and five-membered ring (*furanose*) structures (Fig. 2-5).

The formation of a hemiacetal ring gives rise to a new chiral center in the aldoses, namely, at the C-1 atom. This leads to two C-1 *epimers,** termed *anomers.* When the hydroxyl group at C-1 (glycosidic hydroxyl) in the Fischer projection formula is located on the same side as the hemiacetal ring, the anomer is termed the α form; in the opposite case it is the β form (Fig. 2-5).

In Haworth's perspective formula the plane of the pyranose or furanose ring is assumed to be perpendicular to the plane of paper. The substituents are parallel to this plane and project either above or below the plane of the ring (Fig. 2-6). Hydrogen atoms are indicated by a bar or not at all.

Another system for depicting sugar structures has been proposed by Mills (Fig. 2-7). In Mills' formulas the ring is parallel to the plane of the paper and the substituents above the plane of the ring are denoted by heavy lines whereas dotted lines are used for substituents below the plane. In the case of substituents with undefined orientation or, for a mixture consisting of α and β anomers, a wiggly line is used.

2.2.3 Mutarotation

On dissolution of sugars in water, the optical rotation of the solution changes continuously until an equilibrium is reached. This phenomenon,

---

*Diastereoisomers which differ in their configuration only at one of the carbon atoms are termed epimers. To specify the epimers the site is indicated, for example C-2 epimers or C-3 epimers. Generally, if not indicated, the term epimer refers to diastereoisomers having reversed configuration at the asymmetric carbon atom adjacent to the anomeric center.

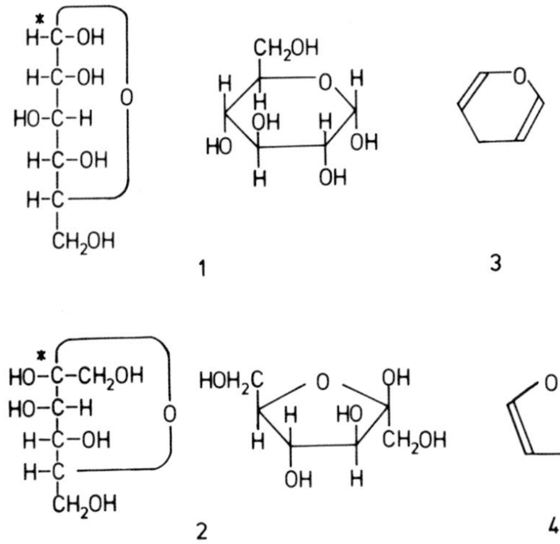

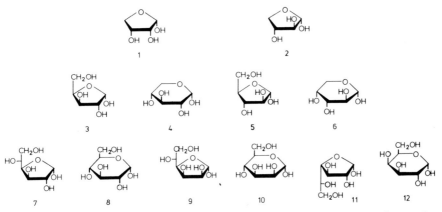

Fig. 2-5. 1, α-D-Glucopyranose (six-membered ring); 2, β-D-fructofuranose (five-membered ring); 3, pyran; 4, furan.

termed *mutarotation,* is accompanied by complex changes as the hemiacetal ring is opened and products with furanose and pyranose rings are formed, which in addition can either have an α or β anomeric configuration (Fig. 2-8). Some of the isomerization reactions are faster than others, and thus a certain form may intermittently reach a high concentration, although it may be only a minor component after the equilibrium has been reached.

Fig. 2-6. Cyclic forms of some α-D-aldoses (Haworth formulas). 1, Erythrose; 2, threose; 3, xylofuranose; 4, xylopyranose; 5, arabinofuranose; 6, arabinopyranose; 7, glucofuranose; 8, glucopyranose; 9, mannofuranose; 10, mannopyranose; 11, galactofuranose; 12, galactopyranose.

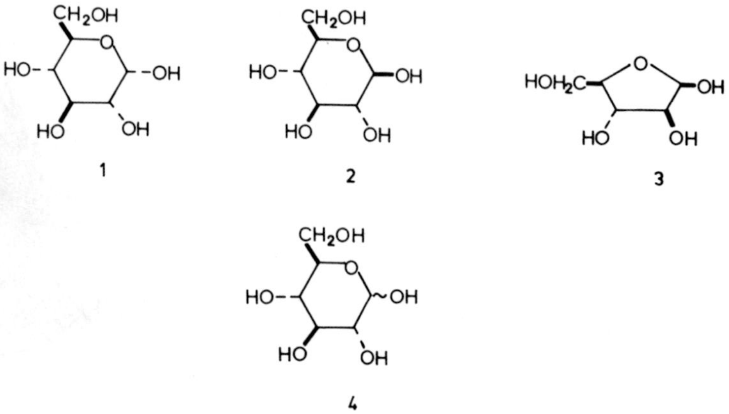

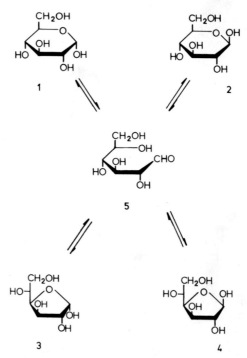

Fig. 2-7. Mills formulas. 1, α-D-Glucopyranose; 2, β-D-glucopyranose; 3, β-D-arabinofuranose; 4, α,β-D-glucopyranose.

Fig. 2-8. Mutarotation of D-glucose. 1, α-D-Glucopyranose; 2, β-D-glucopyranose; 3, α-D-glucofuranose; 4, β-D-glucofuranose; 5, the open-chain aldehyde form.

The final proportions of the four possible ring isomers vary considerably among different sugars, depending on their thermodynamic stabilities. The equilibrium is also affected by the solvent; for instance, the proportion of the furanose form is increased in dimethyl sulfoxide since the solvation of hydroxyl groups is decreased. Traces of acids or bases accelerate this interconversion.

2.2.4   The Conformations of Monosaccharides

Any molecule of a given configuration can exist in different spatial arrangements (*conformations*) when the atoms or atomic groups are rotated or twisted with respect to each other within the limits permitted by the bonds. Although the concept of conformation in carbohydrate chemistry is old (Haworth, 1929), novel studies during the last three decades, especially by Barton and Hassel (Nobel Prize in 1969), have added clarity and important details to this concept. The conformations can best be visualized with the use of molecular models.

The conformations of the six-membered ring systems are better characterized than those of the less stable five-membered analogues. For example, the cyclohexane molecule can occur in two strainless forms, namely in the rigid *chair* form or in the flexible form (Fig. 2-9). The latter can exist in a variety of shapes of which only the *boat* and the *skew boat* (or *twist*) are regular and easily depictable on paper. The chair form is preferred energetically because it is usually free from steric interactions whereas the flexible forms are not. The *half-chair* conformation is possible, when a six-membered ring contains either a double bond or an oxiran ring. In the half-chair conformation four adjacent atoms are in the same plane.

In both the chair and the boat forms of the cyclohexane molecule the

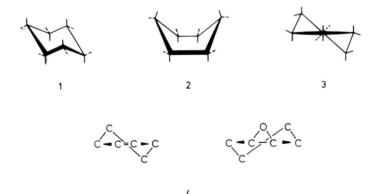

Fig. 2-9.   Conformations of cyclohexane. 1, Chair; 2, boat; 3, skew boat; 4, half-chair.

Fig. 2-10. Ring inversion of a monosubstituted cyclohexane derivative; equatorial substituents become axial and vice versa.

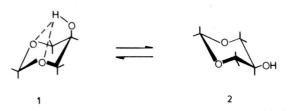

Fig. 2-11. Chair conformations of β-D-glucopyranose. Form 1, in which the OH and CH_2OH substituents are equatorial, is favored.

Fig. 2-12. Hydrogen-bonding interactions. Form 1 is more stable than 2.

Fig. 2-13. Nomenclature system for the chair forms. The conformations 1 and 2 are marked by the symbols $^4C_1$ (C1) and $^1C_4$ (1C). The shadowed area indicates the reference plane.

bonds bearing substituents (H atoms) are oriented either *equatorially* or *axially*. The axial substituents (a) become equatorial (e), and vice versa, when the conformation of the molecule is switched between the two possible chair forms as shown in Fig. 2-10. In monosubstituted molecules the substituent favors the equatorial position as a consequence of minimum nonbonded interaction with the neighboring hydrogen atoms. This is also generally the case for derivatives with several substituents, and the molecule usually takes the conformation in which the majority of substituents are equatorial. When the above rules are applied to D-glucopyranose, its conformation can be presented according to Fig. 2-11.

In some cases the hydroxyl groups are oriented axially, for instance in molecules bearing ring oxygen atoms which participate in hydrogen bonds with the hydroxyl groups (Fig. 2-12). An exception to the tendency of the hydroxyl groups to be equatorially oriented in a pyranose ring is the substituent bound to the anomeric center where an axial position is favored (the so-called *anomeric effect*). The anomeric effect depends on the nature of the substituent and is especially strong for halogens.

The conformations are often denoted as suggested by Reeves (C1 or 1C), but this system is applicable only to the chair form. In a more modern system the form is given by initial letters, i.e., C for the chair, B for the boat, etc., and the two substituents deviating from the reference plane are then marked by two number indexes as shown in Fig. 2-13.

2.3 Monosaccharide Derivatives

Sugar derivatives are, in principle, formed (1) by reaction of the free carbonyl group or the anomeric hydroxyl at C-1 or (2) by reactions of the other hydroxyl groups. For reactions of free carbonyl groups, see Section 2.5.3.

2.3.1 Glycosides

Sugars react as *hemiacetals* with hydroxyl compounds, such as alcohols and phenols forming glycosides (Fig. 2-14). Glycosides exist either as pyranosides or furanosides, each of which possesses two anomeric forms (α and β glycosides). The group derived from the hydroxyl compound is termed the *aglycone*. Figure 2-15 gives examples of the preparation and structure of glycosides. The glycosides are easily hydrolyzed by acids to sugars and alcohols but are usually fairly stable toward alkali.

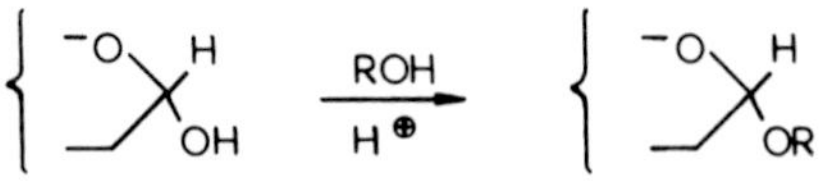

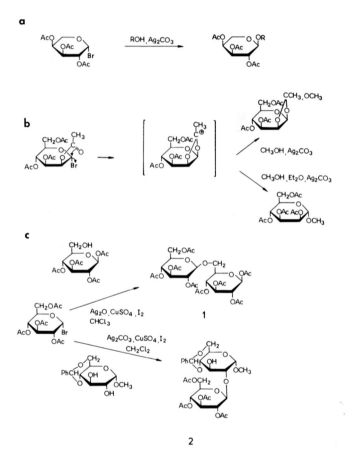

Fig. 2-14.   Glycoside formation. Group R is the aglycone.

Fig. 2-15.   Preparation methods of glycosides (the Koenigs–Knorr synthesis). (a) Reaction of 2,3,4-tri-O-acetyl-β-L-arabinosyl bromide with an alcohol in the presence of silver carbonate. When the acetyl groups at C-2 and the bromide at C-1 are in *cis* position, the displacement proceeds by an unimolecular mechanism resulting in a *trans*-1,2-glycoside. (b) 1,2-*trans* configurations (e.g., 2,3,4,6-tetra-O-acetyl-α-D-mannosyl bromide) give orthoesters in the presence of methanol and silver carbonate. If diethyl ether is added as a diluent, the major products are *trans*-1,2-glycosides. (c) Example of a disaccharide synthesis. The yield is higher when a primary hydroxyl group, as in (1) (β-gentiobiose octaacetate), participates instead of a secondary hydroxyl group, as in (2) (a sophrose derivative), which is sterically hindered.

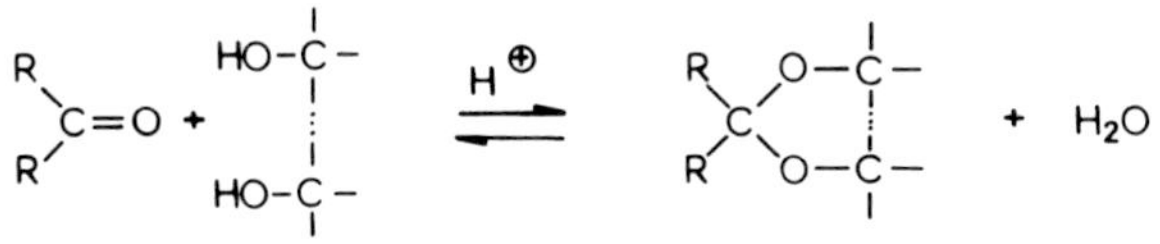

$$\underset{R'}{\overset{R}{\diagdown}}C=O + \begin{array}{c} HO-\overset{|}{C}- \\ \vdots \\ HO-\overset{|}{C}- \end{array} \xrightleftharpoons[]{H^{\oplus}} \begin{array}{c} R \diagdown \; O-\overset{|}{C}- \\ R' \diagup C \diagdown \; O-\overset{|}{C}- \end{array} + H_2O$$

Fig. 2-16. Acetal formation.

2.3.2 Acetals

The most typical acetals are those in which the carbonyl group belongs to a nonsugar component (Fig. 2-16), although the dimethyl acetal of glucose represents another type. Some examples of acetals are given in Fig. 2-17.

2.3.3 Ethers

Ethers are important derivatives of both monosaccharides and polysaccharides. Etherification is often used in the determination of structures and types of linkages between sugars in oligo- and polysaccharides. Table 2-2 gives examples of the preparation of ethers. Ethers are very stable against both acids and bases.

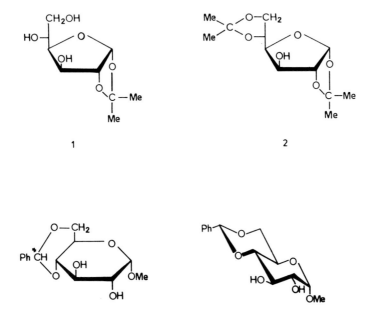

Fig. 2-17. Examples of acetals. 1, 1,2-O-Isopropylidene-α-D-glucofuranose; 2, 1,2:5,6-di-O-isopropylidene-α-D-glucofuranose; 3, methyl-4,6-O-benzylidene-α-D-glucopyranoside.

TABLE 2-2. Some Preparation Methods of Ethers

Ether		Reagent
Methyl	$(ROCH_3)$	a. CH_3I + NaH + DMF
		b. $(CH_3)_2SO_4$ + NaOH
		c. CH_2N_2 + $BF_3 \cdot Et_2O$
Trimethylsilyl	$(ROSi(CH_3)_3)$	$(CH_3)_3SiNHSi(CH_3)_3$ +
		$(CH_3)_3SiCl$ + pyridine
Triphenylmethyl	$(ROC(C_6H_5)_3)$	$(C_6H_5)_3CCl$ + pyridine
(trityl)		

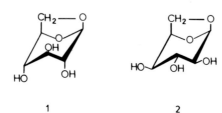

1   2

Fig. 2-18. Examples of anhydro sugars. 1, 1,6-Anhydro-β-D-glucopyranose (levoglucosan); 2, 1,6-anhydro-β-D-idopyranose. In 1,6-anhydro derivatives the groups at C-1 and C-5 are both axial.

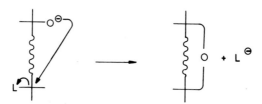

Fig. 2-19. Internal ether (epoxide) formation. Inversion occurs at the carbon atom involved, if the leaving group L is not at a primary position.

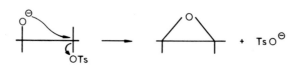

$$Ts = p - CH_3C_6H_4SO_2$$

Fig. 2-20. Oxiran (three-membered ether) formation.

2.3.4 Anhydro Sugars

Glycosans   Anhydro sugars are formed from sugars by the elimination of water from a pair of hydroxyl groups. Glycosans are strictly intramolecular glycosides. In these derivatives the anomeric hydroxyl is involved in the formation of the anhydro linkage. These linkages are readily opened by action of acids, some of them also by bases. The 1,6-anhydroaldohexoses are the most common glycosans (Fig. 2-18).

Epoxides   According to the definitions the internal ethers are derived only from alcoholic hydroxyls and the hydroxyl group in the anomeric center does not participate. They are formed when the sugar molecule contains both a good "leaving group" and a suitably located ionized hydroxyl group.

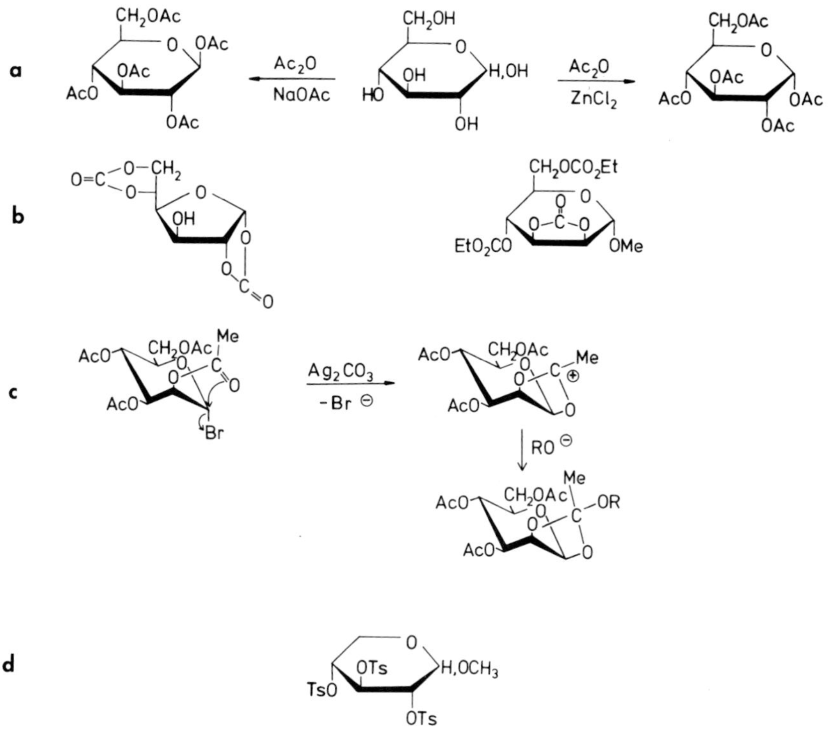

Fig. 2-21.   Some sugar esters. (a) Preparation of D-glucopyranose pentaacetate. The preference of α or β anomers depends on the reaction catalyst. (b) Examples of sugar carbonates prepared by using phosgene or ethyl chloroformate as reagents. (c) Formation of orthoesters (cf. Fig. 2-15). (d) Sugar sulfonates. The most important sulfonyl esters are p-toluenesulfonate or tosylate (OTs) and methanesulfonate or mesylate (OMs). They are prepared from the corresponding sulfonyl chlorides (RSO$_2$Cl). Sulfonyl esters are excellent leaving groups in nucleophilic reactions via which a hydroxyl group can be replaced by some other functional groups.

Since this reaction proceeds according to the S_N2 mechanism, inversion takes place at the carbon atom involved (Fig. 2-19). The ring size of epoxides can vary from three- to six-membered rings. The three-membered derivatives belong to the most important subclass of internal ethers and are termed *oxirans*. A prerequisite for oxiran formation is obviously coplanarity and *trans* position of the reacting groups (Fig. 2-20).

2.3.5   Esters

Sugar esters differ from ethers by being readily hydrolyzed by alkali. Methods for the esterification of polyhydroxy compounds do not generally differ from those applicable for the esterification of simple alcohols. Primary hydroxyls may in certain cases be esterified selectively, since they are more reactive than the secondary hydroxyls. Sometimes a partial esterification of the hydroxyls is desirable. The preparation of esters is clarified by examples collected in Fig. 2-21.

2.4   Oligo- and Polysaccharides

More than 500 oligosaccharides are known today, most of them occurring as free natural substances. Oligosaccharides are also obtained by partial

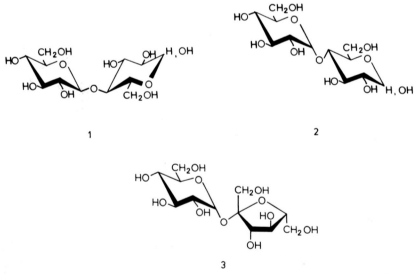

Fig. 2-22.   1, 4-O-(β-D-Glucopyranosyl)-D-glucopyranose (cellobiose); 2, 4-O-(α-D-glucopy-ranosyl)-D-glucopyranose (maltose); 3, α-D-glucopyranosyl β-D-fructofuranoside (sucrose).

acidic or enzymic hydrolysis of polysaccharides. Disaccharides can be considered to be glycosides in which the aglycone part is another monosaccharide. Disaccharides are called *reducing* or *nonreducing*, depending on whether one or both reducing groups are involved in the formation of the glycosidic linkage. *Cellobiose* and *maltose* obtained by partial hydrolysis of cellulose and starch, respectively, are reducing disaccharides (Fig. 2-22). The nonreducing type is exemplified by *sucrose*, which is the most important disaccharide occurring in plants. A large number of various oligosaccharides, up to hexasaccharides, are known.

Polysaccharides are the most abundant constituents of living matter. They are in principle built up in the same manner as oligosaccharides. The chain molecules can be either linear or branched, a fact that markedly affects the physical properties of the polysaccharides. The carbohydrate material in plants is largely composed of cellulose and hemicelluloses. Chapter 3 deals with their structure and properties.

2.5 The Reactions of Carbohydrates

Many reactions of wood polysaccharides (cellulose and hemicelluloses) are described in connection with pulping chemistry (Chapters 7 and 8). The following is therefore restricted to the most important and typical reactions.

2.5.1 Oxidation

By mild oxidants, e.g., aqueous bromine, aldoses (or aldehyde end groups in oligo- and polysaccharides) are oxidized to *aldonic acids* or to corresponding aldonic acid end groups (Fig. 2-23), whereas ketoses are resistant. Stronger oxidants, e.g., nitric acid, convert aldoses to dicarboxylic acids, termed *aldaric acids*. Aldonic and aldaric acids occur in acidic solution mainly in the form of lactones which are intramolecular esters. Exclusive oxidation of the primary carbon atom (C-6 in aldohexoses) to a carboxyl group, which can be accomplished using blocking groups, gives *uronic acids* (Fig. 2-24). Uronic acids are important consituents in wood polysaccharides (see Section 3.3).

Uloses are derivatives of carbohydrates, which contain a further keto group. Aldosuloses are obtained from aldoses and diuloses from ketoses. They are important intermediates in the synthesis of carbohydrates. Uloses can be prepared by oxidation of derivatives in which all the hydroxyls except that one subjected to oxidation are blocked. Uloses are formed as intermediates during pulp bleaching (see Section 8.1.3).

Periodic acid is a specific oxidant for any combination of hydroxyl, car-

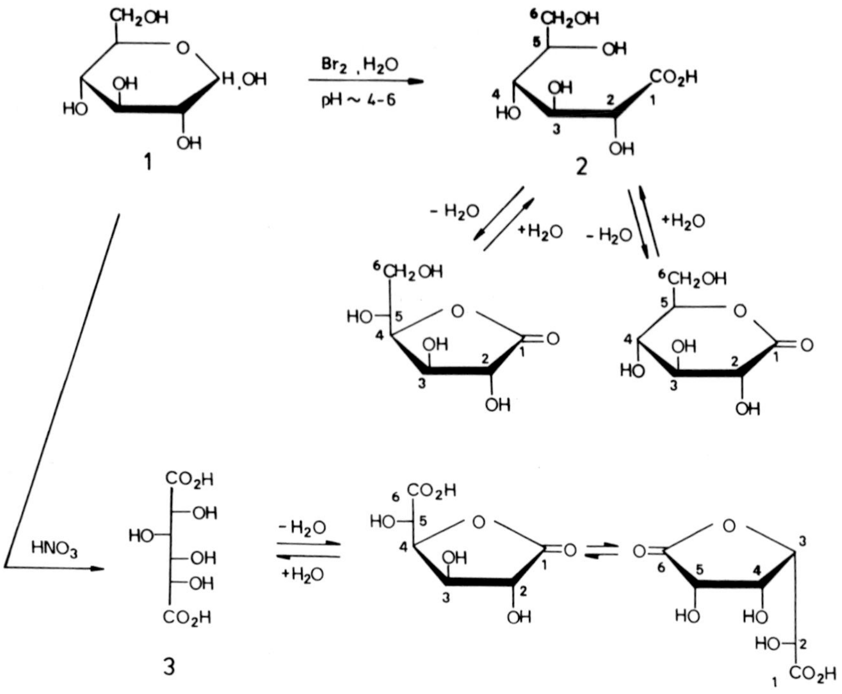

Fig. 2-23.   Oxidation of D-glucose (1) to D-gluconic acid (2) and D-glucaric acid (3). The free acids occur mainly as corresponding lactones.

bonyl, or primary amine groups attached to adjacent carbon atoms (Fig. 2-25). Primary alcoholic groups are oxidized to formaldehyde, secondary to higher aldehydes, and tertiary to ketones. α-Hydroxyaldehydes are oxidized to formic acid and an aldehyde. Since this specific reaction proceeds quantitatively, it is extremely useful for structural studies (see also Section 2.6).

2.5.2   Reduction

Aldoses and ketoses can be reduced to alditols by various agents for which purpose sodium borohydride is very useful. For industrial production of alditols, however, electrolytic reduction is applied. Only one product is formed from aldoses, whereas ketoses give rise to two diastereoisomers because of the generation of a new asymmetric center (Fig. 2-26). Sodium borohydride can also be used for reduction of carbonyl groups in polysaccharides.

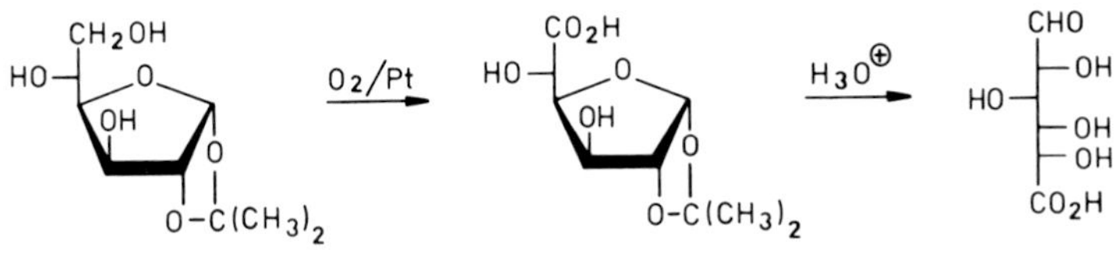

Fig. 2-24. Preparation of D-glucuronic acid from D-glucose. The primary hydroxyl group at C-6 is selectively oxidized after protection of the anomeric center (see Fig. 2-17). The most useful reagent for oxidation is oxygen in the presence of platinum metal but potassium permanganate and dinitrogen tetroxide can also be applied.

2.5.3 Addition and Condensation Reactions of Carbonyl Groups

In the classic carbohydrate chemistry addition reactions of carbonyl groups served as valuable tools for sturctural studies of carbohydrates. For example, hydroxylamine, hydrazine, and phenylhydrazine react with carbonyl groups to yield oximes and hydrazones. In the presence of an excess

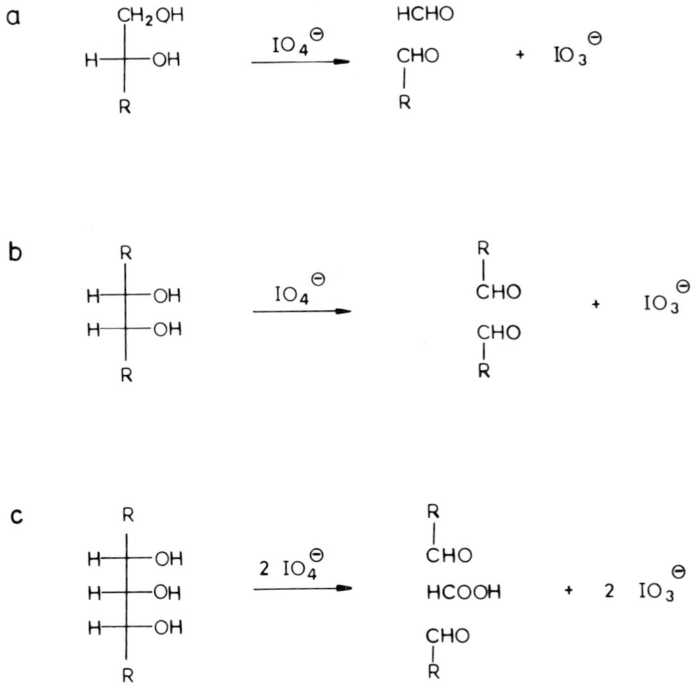

Fig. 2-25. Periodate oxidation of polyhydroxy compounds.

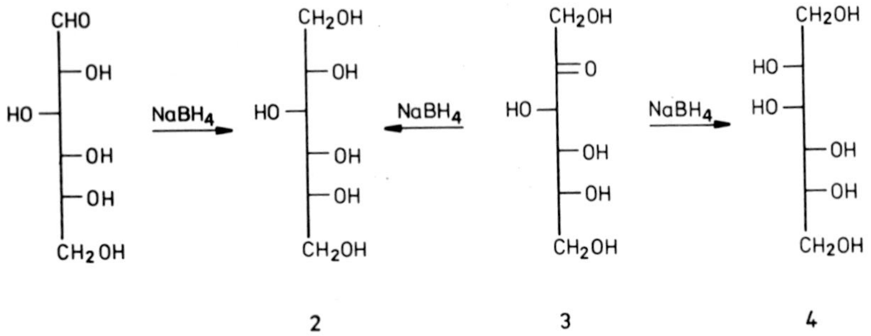

$$2 \qquad 3 \qquad 4$$

Fig. 2-26.   Reduction of aldoses and ketoses. D-Glucose (1) yields only D-glucitol (2), which also is formed from D-fructose (3) in addition to D-mannitol (4).

of phenylhydrazine, the C-2 position is oxidized to a carbonyl group and a *phenylosazone derivative* is formed (Fig. 2-27). Sugars, which differ only at C-1 and C-2 positions, for example, glucose, mannose, and fructose, give the same osazone (Fischer). Although mainly spectroscopic methods and chromatography are applied today for structural studies and identification purposes, these reagents are useful for simple identification and for determination of carbonyl groups in oxidized cellulose samples.

Cyanide ions react with aldehydes and ketones to yield *cyanohydrins* (Kiliani) (Fig. 2-28). Hydrolysis of the cyanohydrins gives aldonic acids, which can be reduced to aldoses. Kiliani reaction thus opens the possibility for chain lengthening of aldoses. Because of the formation of a hydroxyl group in place of the aldehyde group a new asymmetric center is generated. It is to be observed, however, that the reaction is subject to so-called *asymmetric induction*, which means that the diastereoisomers are formed in unequal proportions.

Another type of addition reaction is represented by the reaction of hydro-

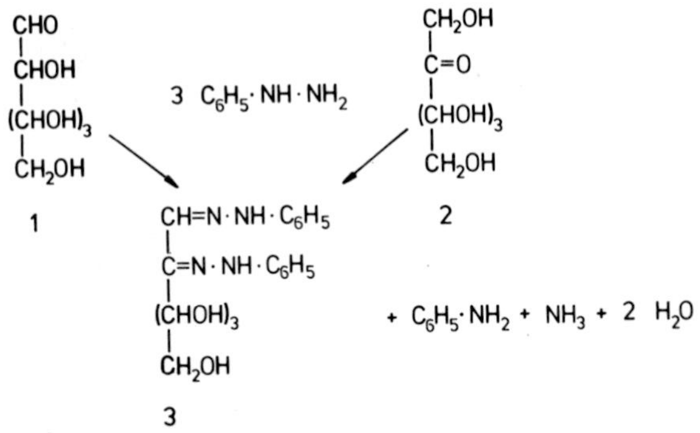

Fig. 2-27.   Formation of phenylosazone. 1, Aldose; 2, ketose; 3, phenylosazone.

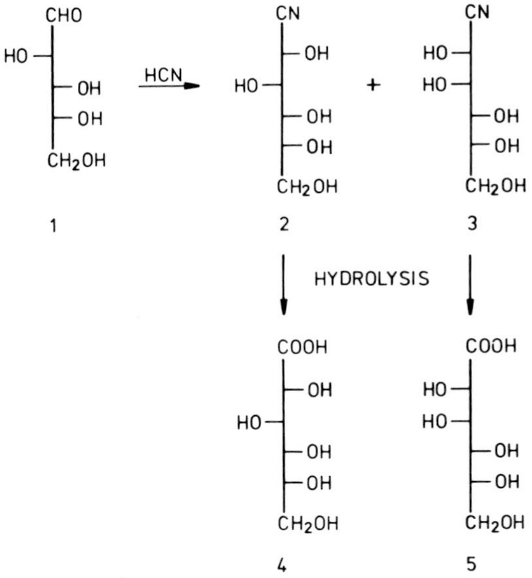

CHO
HO —
|— OH
|— OH
CH₂OH

1

→ HCN →

CN
|— OH
HO —
|— OH
|— OH
CH₂OH

2

+

CN
HO —
HO —
|— OH
|— OH
CH₂OH

3

HYDROLYSIS

COOH
|— OH
HO —
|— OH
|— OH
CH₂OH

4

COOH
HO —
HO —
|— OH
|— OH
CH₂OH

5

Fig. 2-28. Kiliani reaction. Addition of hydrogen cyanide to D-arabinose (1) yields cyanohydrins (2) and (3), which are hydrolyzed to D-gluconic acid (4) and D-mannonic acid (5). Because of asymmetric induction preferentially (4) is formed.

gen sulfite ions with sugars giving rise to the formation of α-hydroxysulfonic acids (Fig. 2-29). The equilibrium of this reaction depends on the configuration of the sugar; for example, mannose and xylose form more stable bisulfite addition products than glucose, and ketoses (fructose) show almost negligible affinity toward hydrogen sulfite ions. The bisulfite addition reaction has been applied for separation of monosaccharides. Sulfite spent liquors contain so-called loosely combined sulfur dioxide bound to sugars and other carbonyl-bearing constituents (see Section 7.2.9).

2.5.4 The Influence of Acid

The acidic hydrolysis of glycosidic bonds is of importance in many technical processes based on wood as raw material. Figure 2-30 illustrates the

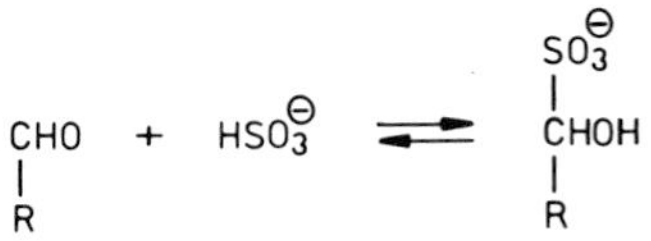

$$
\begin{array}{ccc}
\text{CHO} & & \text{SO}_3^{\ominus} \\
| & + \ \text{HSO}_3^{\ominus} \ \rightleftharpoons & | \\
\text{R} & & \text{CHOH} \\
& & | \\
& & \text{R}
\end{array}
$$

Fig. 2-29. Formation of α-hydroxysulfonic acids after addition of hydrogen sulfite ions to aldoses. R = the monosaccharide residue.

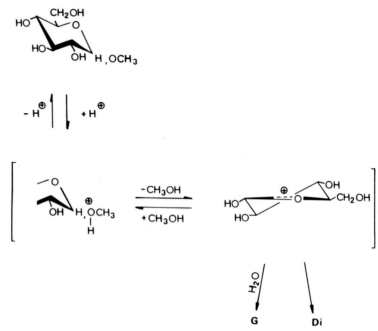

Fig. 2-30. Acid-catalyzed hydrolysis of glucopyranosides. Besides glucose (G) also small amounts of disaccharides (Di) are formed.

mechanism of the acidic cleavage of glycosidic bonds. The reaction starts with a rapid proton addition to the aglycon oxygen atom followed by a slow breakdown of the protolyzed conjugate acid to the cyclic carbonium ion, which adopts a half-chair conformation. After a rapid addition of water, free sugar is liberated. Because the sugar competes with the solvent (water) small amounts of disaccharides are formed as *reversion* products.

The rate of hydrolysis of polysaccharides is affected by several factors. Because of substituent interaction effects, furanosides are hydrolyzed much more rapidly than the pyranoside analogues. Differences in the hydrolysis rates of diastereomeric glycosides are significant. For example, the relative hydrolysis rates of methyl-α-D-gluco-, manno-, and galactopyranosides are 1.0:2.9:5.0. This can be related to the stabilities of the respective conjugate acids, which are transformed into the half-chair carbonium ions at different rates. Also, substituents bound to the C-2 position obviously prevent the formation of the half-chair conformation.

Carboxyl groups bound to the polysaccharide chains have a considerable influence on the rate of acid hydrolysis probably mainly because of steric interaction even if inductive effects should also be considered. For example, glycuronides are hydrolyzed more slowly than glycosides. It can be assumed that the formation of the intermediate carbonium ion takes place more rapidly at the end than in the middle of the polysaccharide chain. In accordance with this the yield of monosaccharides after partial hydrolysis or sulfite pulping is higher than calculated on the basis of a random bond cleavage.

An opposite reaction to the acid-catalyzed hydrolysis is the above-mentioned reversion. Acids can also catalyze the formation of anhydro sugars (see Section 2.3.4). Reversion tends to result in formation of (1→6)-glycosidic bonds. The degradation of pentoses and uronic acids into furfural and of hexoses into hydroxymethylfurfural, levulinic, and formic acids are also important acid-catalyzed reactions, which, however, require concentrated acid and higher temperatures (Fig. 2-31).

2.5.5   The Influence of Alkali

In weakly alkaline solutions aldoses and ketoses undergo rearrangements. An example is the *Lobry de Bruyn –Alberda van Ekenstein* transformation of

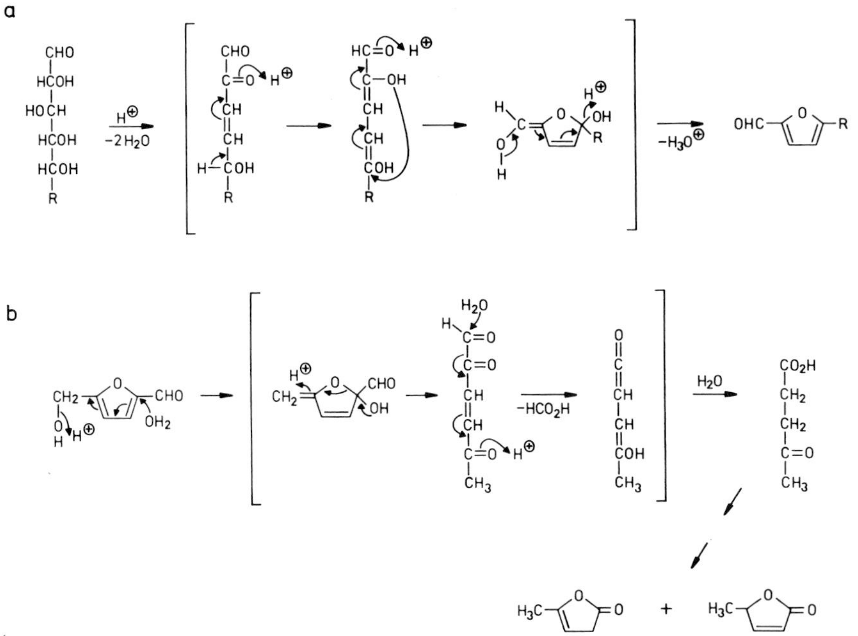

Fig. 2-31. Reactions of sugars in the presence of concentrated mineral acids. (a) Pentoses (R = H) yield furfural and hexoses (R = CH_2OH) hydroxymethylfurfural. (b) On further heating hydroxymethylfurfural is fragmented under liberation of formic acid. The rest of the molecule is rearranged to levulinic acid, which is lactonized to form α- and β-angelica lactones.

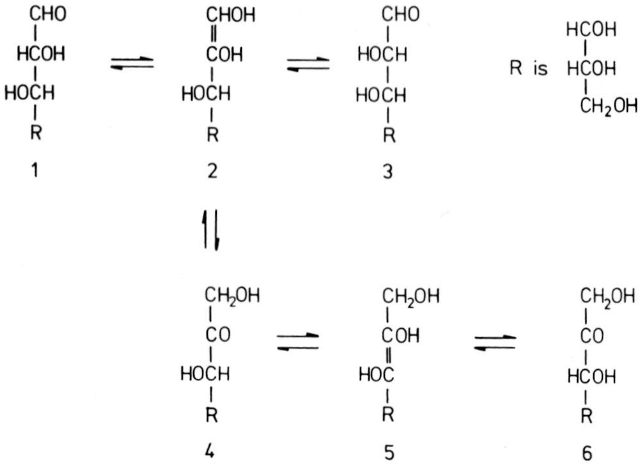

$$
\begin{array}{c}
\text{CHO} \\
| \\
\text{HCOH} \\
| \\
\text{HOCH} \\
| \\
\text{R} \\
\mathbf{1}
\end{array}
\rightleftharpoons
\begin{array}{c}
\text{CHOH} \\
\| \\
\text{COH} \\
| \\
\text{HOCH} \\
| \\
\text{R} \\
\mathbf{2}
\end{array}
\rightleftharpoons
\begin{array}{c}
\text{CHO} \\
| \\
\text{HOCH} \\
| \\
\text{HOCH} \\
| \\
\text{R} \\
\mathbf{3}
\end{array}
\qquad
\text{R is}
\begin{array}{c}
\text{HCOH} \\
| \\
\text{HCOH} \\
| \\
\text{CH}_2\text{OH}
\end{array}
$$

$$
\begin{array}{c}
\text{CH}_2\text{OH} \\
| \\
\text{CO} \\
| \\
\text{HOCH} \\
| \\
\text{R} \\
\mathbf{4}
\end{array}
\rightleftharpoons
\begin{array}{c}
\text{CH}_2\text{OH} \\
| \\
\text{COH} \\
\| \\
\text{HOC} \\
| \\
\text{R} \\
\mathbf{5}
\end{array}
\rightleftharpoons
\begin{array}{c}
\text{CH}_2\text{OH} \\
| \\
\text{CO} \\
| \\
\text{HCOH} \\
| \\
\text{R} \\
\mathbf{6}
\end{array}
$$

Fig. 2-32. Lobry de Bruyn–Alberda van Ekenstein transformation of sugars. 1, D-Glucose; 2, 1,2-enediol; 3, D-mannose; 4, D-fructose; 5, 2,3-enediol; 6, D-allulose.

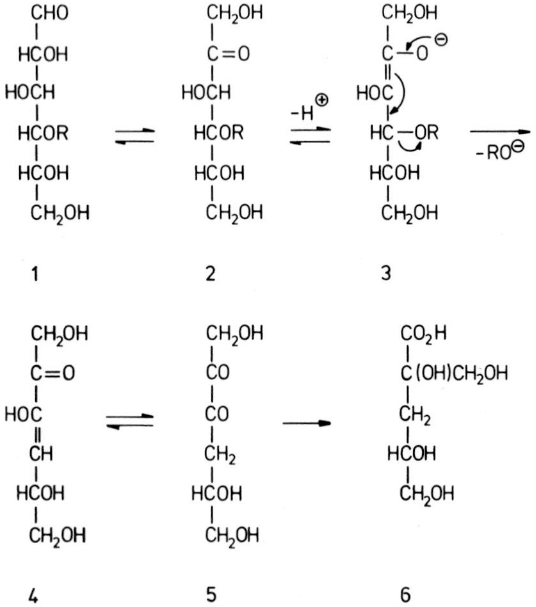

$$
\begin{array}{c}
\text{CHO} \\
| \\
\text{HCOH} \\
| \\
\text{HOCH} \\
| \\
\text{HCOR} \\
| \\
\text{HCOH} \\
| \\
\text{CH}_2\text{OH} \\
\mathbf{1}
\end{array}
\rightleftharpoons
\begin{array}{c}
\text{CH}_2\text{OH} \\
| \\
\text{C}{=}\text{O} \\
| \\
\text{HOCH} \\
| \\
\text{HCOR} \\
| \\
\text{HCOH} \\
| \\
\text{CH}_2\text{OH} \\
\mathbf{2}
\end{array}
\xrightarrow[\text{-H}^{\oplus}]{}
\begin{array}{c}
\text{CH}_2\text{OH} \\
| \\
\text{C}{-}\text{O}^{\ominus} \\
\| \\
\text{HOC} \\
| \\
\text{HC}{-}\text{OR} \\
| \\
\text{HCOH} \\
| \\
\text{CH}_2\text{OH} \\
\mathbf{3}
\end{array}
\xrightarrow{-\text{RO}^{\ominus}}
$$

$$
\begin{array}{c}
\text{CH}_2\text{OH} \\
| \\
\text{C}{=}\text{O} \\
| \\
\text{HOC} \\
\| \\
\text{CH} \\
| \\
\text{HCOH} \\
| \\
\text{CH}_2\text{OH} \\
\mathbf{4}
\end{array}
\rightleftharpoons
\begin{array}{c}
\text{CH}_2\text{OH} \\
| \\
\text{CO} \\
| \\
\text{CO} \\
| \\
\text{CH}_2 \\
| \\
\text{HCOH} \\
| \\
\text{CH}_2\text{OH} \\
\mathbf{5}
\end{array}
\longrightarrow
\begin{array}{c}
\text{CO}_2\text{H} \\
| \\
\text{C(OH)CH}_2\text{OH} \\
| \\
\text{CH}_2 \\
| \\
\text{HCOH} \\
| \\
\text{CH}_2\text{OH} \\
\mathbf{6}
\end{array}
$$

Fig. 2-33. Alkaline peeling reaction of cellulose (R = cellulose chain). 1 → 2, Isomerization; 2 → 3, 2,3-enediol formation; 3 → 4, β-alkoxy elimination; 4 → 5, tautomerization; 5 → 6, benzilic acid rearrangement leading to glucoisosaccharinic acid.

aldoses (Fig. 2-32). This reaction starts by enolization of the aldose to an 1,2-enediol which can be converted to either of the two C-2 epimeric aldoses or to a ketose which also can undergo epimerization. Also aldonic acids are epimerized by alkali, especially in pyridine solutions.

Strong alkali converts monosaccharides, as well as the end groups in polysaccharides, to various carboxylic acids. (1→4)-Linked polysaccharides including cellulose and most hemicelluloses, are degraded by an endwise mechanism, known as the peeling reaction. This reaction occurs during alkaline pulping and bleaching processes, for example in kraft pulping and oxygen bleaching (see Sections 7.3.5 and 8.1.3). The reaction mechanism is outlined in Fig. 2-33. The degradation starts with the isomerization of the end group to a ketose in which the glycosidic bond is in the β position with respect to the carbonyl group. Since such a structure is labile in alkali, the glycosidic bond is cleaved with removal of the end group. This is termed "β-alkoxy elimination." The eliminated end group is tautomerized to a dicarbonyl derivative which then undergoes benzilic acid rearrangement to isosaccharinic acid. In addition, a number of other acids are formed by competing mechanisms. In kraft pulping, the

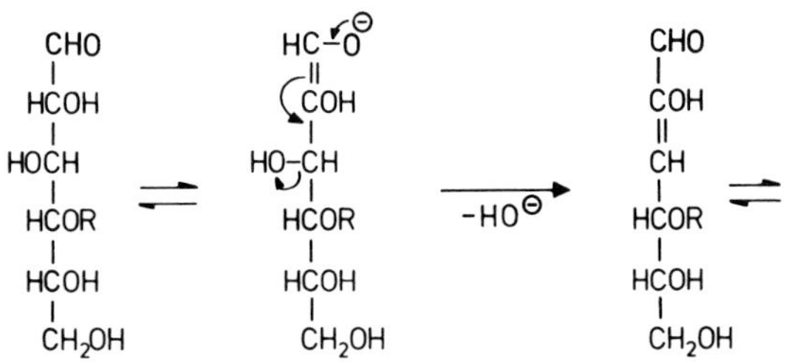

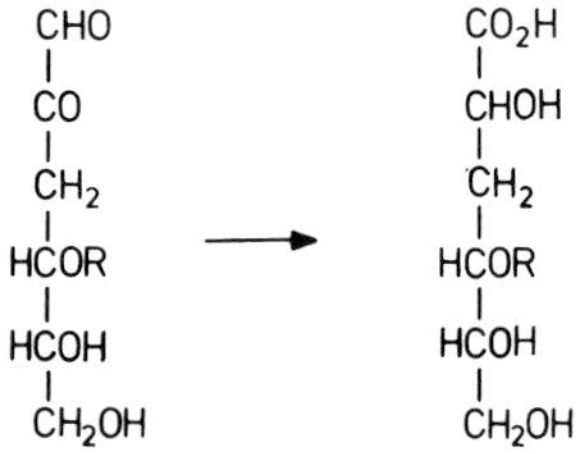

Fig. 2-34. Stopping reaction. 1 → 2, 1,2-Enediol formation; 2 → 3, β-hydroxy elimination; 3 → 4, tautomerization; 4 → 5, benzilic acid rearrangement leading to a glucometasaccharinic acid end group. (cf. Fig. 2-33.)

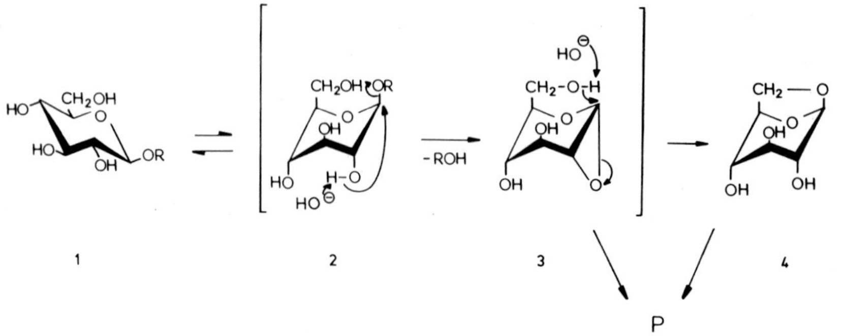

Fig. 2-35. Base-catalyzed hydrolysis of β-D-glucopyranosides (R is a carbohydrate residue or another substituent). The reaction starts with reorientation of the equatorial hydroxyl groups to axial positions (1 → 2 or D-$^4C_1$ → D-$^1C_4$). After formation of an oxiran ring (3) the glycosidic linkage is cleaved leading to different decomposition products (P) directly or via the levoglucosan intermediate (4).

cellulose molecules are subjected to this endwise peeling, which results in a loss of about fifty glucose units from a single molecule. The peeling process is terminated by a so-called *stopping* reaction involving a direct *β-hydroxy elimination* from the C-3 position (Fig. 2-34). The end group undergoes a benzilic acid rearrangement to an alkali-stable metasaccharinic acid end group. Other end groups are, however, also formed. The 3-O-substituted glycosides of (1→4)-linked polysaccharides are rapidly stabilized in alkali through benzilic acid rearrangement because the β-alkoxy elimination takes place much easier than the β-hydroxy elimination.

The cleavage of glycosidic bonds by alkali is usually extremely slow in comparison with the acid-catalyzed hydrolysis. A suggested mechanism for this reaction is depicted in Fig. 2-35. Ionization of the C-2 hydroxyl group and conformational change results in the formation of a three-membered epoxide (oxiran) ring under simultaneous cleavage of the glycosidic bond (elimination of the alkoxy group). The opening of the oxirane ring results in the formation of a free reducing end group in the polysaccharide chain (or free sugar), or, if the steric requirements are fulfilled, a 1,6-anhydride. The mechanism explains why 1,2-*trans*-glycosides are more reactive than the 1,2-*cis*-anomers.

2.6 Structural Studies

A number of novel methods, including those based on spectroscopy and chromatography, have facilitated the structural studies on oligo- and polysaccharides. Some principles of these methods are outlined in the following.

2.6.1 Identification of the Monosaccharide Units

Monosaccharides liberated on acidic hydrolysis of polysaccharides can be identified and determined by combining gas-liquid chromatography (GLC) and mass spectrometry (MS). For this purpose the monosaccharides must be transformed into volatile derivatives, such as trimethylsilyl ethers (TMS). For quantitative analysis it is often best first to reduce the monosaccharides to corresponding alditols using sodium borohydride and to separate the fully acetylated derivatives by GLC. In this case the pyranosidic and furanosidic isomers are eliminated and each alditol, resulting from the respective monosaccharide (aldose), gives a single peak in the chromatogram (cf. Section 2.5.2). A variety of other chromatographic methods, particularly high pressure liquid chromatography (HPLC) and thin layer chromatography (TLC), are commonly used techniques. Paper chromatography was formerly widely used but has now been replaced by more sensitive methods.

2.6.2. The Ring Size of Monosaccharide Units and the Position of Linkages

A simple test with Fehling's solution is sufficient to determine whether the carbohydrate is of the reducing or nonreducing type. The etherification, usually methylation, of the free hydroxyl groups followed by hydrolysis and GLC or GLC-MS of the fragments provides information both on the position of the linkages as well as on the ring sizes. Additional evidence can be obtained by selective oxidation methods, including oxidation with periodate or lead tetraacetate.

2.6.3 The Type of Glycosidic Linkage (α or β Form)

Carbohydrates in nature are optically active and polarimetry is widely used in establishing their structure. Measurement of the specific rotation gives information about the linkage type (α or β form) and is also used to follow mutarotation. Nuclear magnetic resonance spectroscopy (NMR) can be used to differentiate between the anomeric protons in the α- or β-pyranose and furanose anomers and their proportions can be measured from the respective peak areas.

Enzymic methods are gaining wider applications. For example, maltase hydrolyzes α-glucosidic linkages, whereas emulsin is specific for β-glucosidic bonds.

References

Ferrier, R. J., and Collins, P. M. (1972). "Monosaccharide Chemistry." Clowes, London.

Guthrie, R. D. (1974). "Guthrie & Honeyman's Introduction to Carbohydrate Chemistry," 4th ed. Oxford Univ. Press (Clarendon), London and New York.

Harmon, R. E. (1979). "Asymmetry in Carbohydrates." Dekker, New York.

Stoddart, J. F. (1971). "Stereochemistry of Carbohydrates." Wiley (Interscience), New York.

Whistler, R. L., Wolfrom, M. L., and BeMiller, J. N., eds. (1962–1980). *Methods Carbohydr. Chem.* **1–8.**

Chapter 3

WOOD POLYSACCHARIDES

3.1 Biosynthesis

As a result of photosynthesis glucose is produced in the foliage of trees from carbon dioxide and water, and then transported in the phloem to the cambial tissues. It is the basic monomer from which the wood polysaccharides are formed through a variety of biosynthetic pathways. Pioneering work by Leloir (Nobel Prize in 1970) led to the discovery of an important sugar nucleotide, UDP-D-glucose, from which cellulose is synthesized. Another important sugar nucleotide, participating in the synthesis of hemicelluloses, is GDP-D-glucose. The nucleoside moieties of these sugar nucleotides are uridine and guanosine (Fig. 3-1). The sugar nucleotides are formed from the corresponding nucleoside triphosphates and glucose phosphate by an enzymic process as illustrated for UDP-D-glucose in Fig. 3-2.

Cellulose is synthesized from UDP-D-glucose the energy content of which is used for the formation of glucosidic bonds in the growing polymer:

$$\text{UDP-D-glucose} + [(1 \rightarrow 4)\text{-}\beta\text{-D-glucosyl}]_n \longrightarrow$$

$$[(1 \rightarrow 4)\text{-}\beta\text{-D-glucosyl}]_{n+1} + \text{UDP}$$

Present concepts of the biosynthesis of cellulose are outlined in Fig. 3-3. In the synthesis of other wood polysaccharides both UDP-D-glucose and

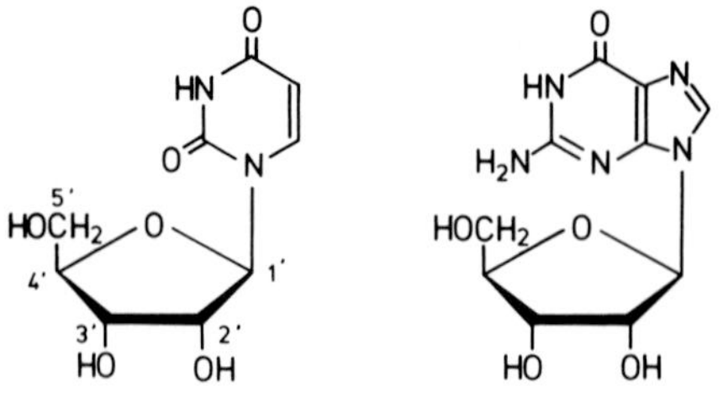

Fig. 3-1. Structure of two nucleosides, 1, Uridine and 2, guanosine, which both contain β-D-ribofuranose residue. The aglycon part is derived from a pyrimidine and purine base, respectively.

GDP-D-glucose are involved, the latter being the principal nucleotide as concerns the formation of mannose-containing hemicelluloses (galactoglucomannans and glucomannans). The monomeric sugar components needed are formed from the nucleotides by complex enzymic reactions involving epimerization, dehydrogenation, and decarboxylation (Fig. 3-4).

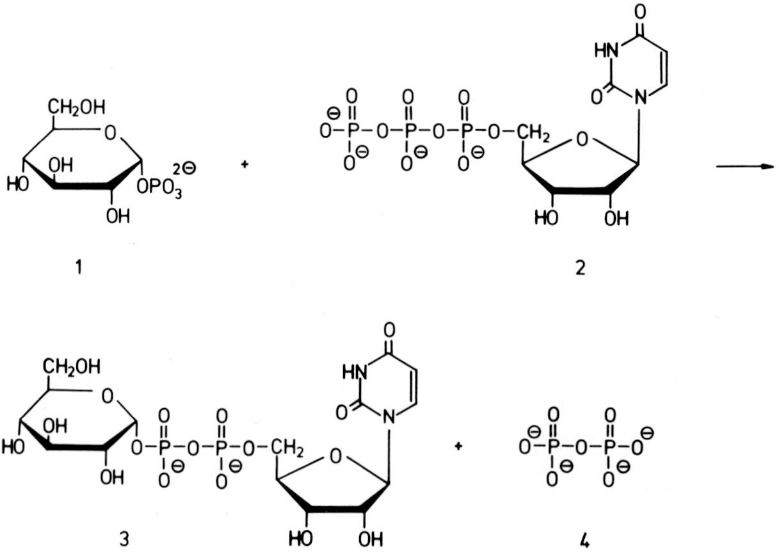

Fig. 3-2. Formation of uridine diphosphate glucose (UDP-D-glucose) (3) from α-D-glucopyranosyl-1-phosphate (1) and uridine triphosphate (2) under simultaneous release of pyrophosphate (4).

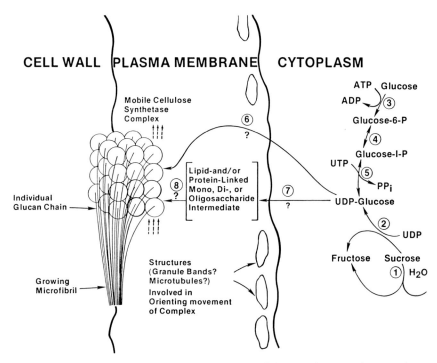

Fig. 3-3. Hypothetical model of the mechanism of cellulose synthesis in plants (Delmer, 1980). UDP-glucose is synthesized in the cytoplasm by cytosolic enzymes. This activated form of glucose may be transferred directly to the growing glucan chains by catalytic sites on subunits of a cellulose synthetase complex embedded in the plasma membrane. Alternatively, an intermediate transfer of glucose residues from UDP-glucose to lipid- and/or protein-linked intermediates may occur prior to transfer to the growing glucan chains of the fibril. The glucan chains derived from one complex are assumed to associate by hydrogen bonding to form a fibril, the size of which may vary among different cell types. As synthesis proceeds, the orientation of the fibrils may be determined by the movement of the complex in the fluid lipid bilayer. Such movement may be directed by microtubules and/or granule bands found on the inner face of the plasma membrane. Numbers refer to reactions catalyzed by the following enzymes: 1: invertase; 2: sucrose synthetase; 3: hexokinase; 4: phosphoglucomutase; 5: UDP-glucose pyrophosphorylase; 6, 7, and 8: hypothetical reactions on the pathway to cellulose. PP_i is pyrophosphate. (Reprinted with permission from The Chemical Rubber Co., CRC Press, Inc.)

3.2   Cellulose

Cellulose is the main constituent of wood. Approximately 40–45% of the dry substance in most wood species is cellulose, located predominantly in the secondary cell wall (cf. Appendix).

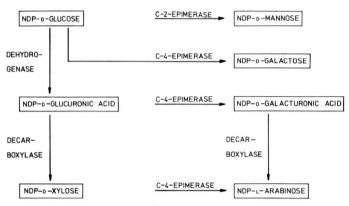

Fig. 3-4. Simplified representation of the formation of hemicellulose precursors from UDP-D-glucose or GDP-D-glucose. Note that NDP (nucleotide diphosphate) means either UDP or GDP.

3.2.1   Molecular Structure

Although the chemical structure of cellulose is understood in detail, its supermolecular state, including its crystalline and fibrillar structure is still open to debate. Examples of incompletely solved problem areas are the exact molecular weight and polydispersity of native cellulose and the dimensions of the microfibrils.

Cellulose is a homopolysaccharide composed of β-D-glucopyranose units which are linked together by (1 → 4)-glycosidic bonds (Fig. 3-5). Cellulose molecules are completely linear and have a strong tendency to form intra- and intermolecular hydrogen bonds. Bundles of cellulose molecules are thus aggregated together in the form of microfibrils, in which highly ordered (crystalline) regions alternate with less ordered (amorphous) regions. Microfibrils build up fibrils and finally cellulose fibers. As a consequence of its fibrous structure and strong hydrogen bonds cellulose has a high tensile strength and is insoluble in most solvents. The physical and chemical behavior of cellulose differs completely from that of starch, which clearly demonstrates the unique influence of stereochemical characteristics. Like cellulose, the amylose component of starch consists of (1 → 4)-linked D-glucopyranose units, but in starch these units are α-anomers. Amylose

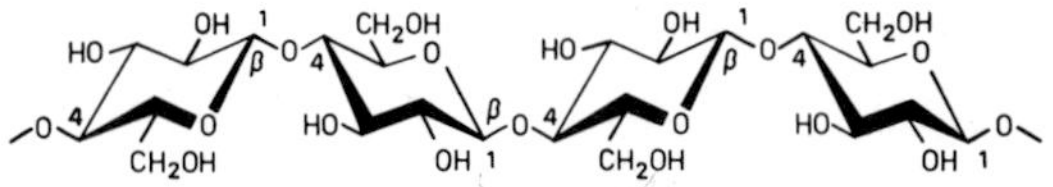

Fig. 3-5. Structure of cellulose. Note that the β-D-glucopyranose chain units are in chair conformation ($^4C_1$) and the substituents HO-2, HO-3, and CH_2OH are oriented equatorially.

occurs as a helix in its solid state and sometimes also in solution. Amylopectin, the other starch component, is also a (1→4)-α-glucan but is highly branched. The branched structure accounts for its extensive solubility, since no aggregation can take place.

The crystalline structure of cellulose has been characterized by X-ray diffraction analysis and by methods based on the absorption of polarized infrared radiation. The unit cell of native cellulose (cellulose I) consists of four glucose residues (Figs. 3-6 and 3-7). In the chain direction (c), the repeating unit is a cellobiose residue (1.03 nm), and every glucose residue is accordingly displaced 180° with respect to its neighbors, giving cellulose a 2-fold axis. It has now been established and largely accepted that all chains in native cellulose microfibrils are oriented in the same direction, that is, they are parallel (Fig. 3-7). There are two hydrogen bonds within each cellulose chain, namely from O(6) in one glucose residue to O(2)H in the adjacent glucose and also from O(3)H to the ring oxygen, as shown in Fig. 3-8. The chains form a layer in the a-c crystallographic plane, where they are held together by hydrogen bonds from O(3) in one chain to O(6)H in the other. There are no hydrogen bonds in cellulose I between these layers, only weak van der Waal's forces in the direction of the b-axis. Native cellulose therefore has a chain lattice and a layer lattice at the same time.

Regenerated cellulose (cellulose II) (Fig. 3-6) has antiparallel chains (Fig. 3-9). The hydrogen bonds within the chains and between the chains in the

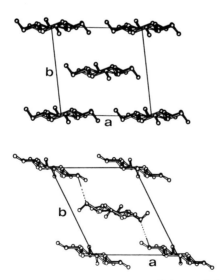

Fig. 3-6. Axial projections of the structures of native cellulose (cellulose I, above) and regenerated cellulose (cellulose II, below). (Reproduced from Kolpak et al., 1978, **19,** 123–131, by permission of the publishers, IPC Business Press Ltd. ©.)

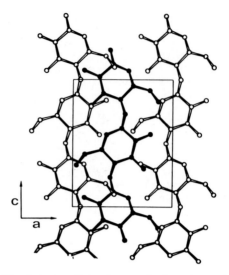

Fig. 3-7. Projection of the chains in cellulose I perpendicular to the ac plane. The center chain (black) is staggered but is parallel with the two corner chains (Gardner and Blackwell, 1974).

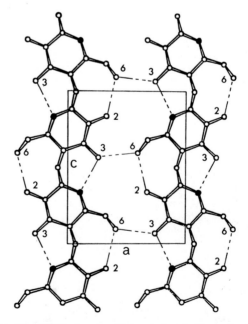

Fig. 3-8. Projection of the (O2O) plane in cellulose I, showing the hydrogen bonding network and the numbering of the atoms. Each glucose residue forms two intramolecular hydrogen bonds (03-H$\cdots$05' and 06$\cdots$H-02') and one intermolecular bond 06-H$\cdots$03). (Slightly modified from Gardner and Blackwell, 1974.)

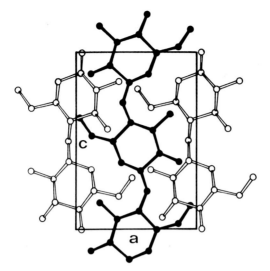

Fig. 3-9. Projection of the chains in cellulose II perpendicular to the *ac* plane. The center chain (black) is staggered and antiparallel to the corner chains. (Reproduced from Kolpak *et al.*, 1978, **19,** 123–131, by permission of the publishers, IPC Business Press Ltd. ©.)

a–*c* plane are the same as in cellulose I. In addition, there are two hydrogen bonds between a corner chain and a center chain (Fig. 3-6), namely from O(2) in one chain to O(2)H in the other and also from O(3)H to 0(6). Cellulose II is formed whenever the lattice of cellulose I is destroyed, for example on swelling with strong alkali or on dissolution of cellulose. Since the strongly hydrogen bonded cellulose II is thermodynamically more stable than cellulose I, it cannot be reconverted into the latter. All naturally occurring cellulose has the structure of cellulose I. Celluloses III and IV are produced when celluloses I and II are subjected to certain chemical treatments and heating.

   The proportions of ordered and disordered regions of cellulose vary considerably depending on the origin of the sample (cf. Table 9-1). Cotton cellulose is more crystalline than cellulose in wood.

3.2.2   The Chain Length and Polydispersity of Cellulose

   The polymer properties of cellulose are usually studied in solution, using solvents, such as CED or Cadoxen (see Section 9.2). On the basis of the solution properties, conclusions can be drawn concerning the average molecular weight, polydispersity, and chain configuration. However, the

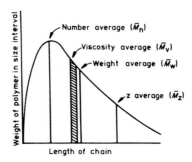

Fig. 3-10. The molecular weight distribution and the average molecular weights of a typical polymer (Billmeyer, 1965).

isolation of cellulose from wood involves risk for some degradation resulting in a reduced molecular weight.

The distribution of molecular weights can be presented statistically as illustrated by Fig. 3-10 where the weight of polymer of a given size is plotted against the chain length. The experimental measurements give an average value of the molecular weight and some methods also a molecular weight distribution. For any polydisperse system, these average values differ from each other depending on the method used. The *number average* molecular weight $\bar{M}_n$ * can be measured using osmometry or by determining the number of reducing end groups. The *weight average* molecular weight $\bar{M}_w$ can be deduced from light scattering data. Sedimentation equilibrium data attainable by ultracentrifugation technique give so-called $\bar{M}_z$ values. Finally, $\bar{M}_v$ refers to the molecular weight calculated on the basis of viscosity measurements. For cellulose, the relationship between molecular weight and degree of polymerization (DP) is DP = M/162, where 162 is the molecular weight of anhydroglucose unit. The ratio $\bar{M}_w/\bar{M}_n$ is a meaure of polydispersity corresponding to the width of the molecular weight distribution and ranges for typical polymers from 1.5–2.0 to 20–50.

Molecular weight measurements have shown that cotton cellulose in its native state consists of about 15,000 and wood cellulose of about 10,000 glucose residues. Some polydispersity data on cellulose derivatives and polysaccharides are shown in Table 3-1. There are indications that the native cellulose present in the secondary cell wall of plants is monodisperse, that is, contains only molecules of one size. In such a case, number and weight average molecular weights ought to be identical. The cellulose in the primary cell wall, on the other hand, which has a lower average molecular

---

*The SI system (Système International d'Unités) recommends the term *relative molecular mass* instead of *molecular weight,* but because the SI term is not yet universally adopted in the polymer chemistry the latter term is used throughout this book.

TABLE 3-1. Polydispersity Values ($\bar{M}_w/\bar{M}_n$) of Different Polysaccharides[a]

Macromolecule	Source	$\bar{M}_w \times 10^{-5}$	$\bar{M}_w/\bar{M}_n$
Cellulose nitrate	Birch	27[b]	1.9[c]
Cellulose nitrate	Ramie	24[b]	1.7[c]
Amylose	Potato	8.8[d]	1.9[d]
Xylan	Birch	0.8[b]	2.3[e]
Xylan	Elm	0.7[b]	2.4[e]
Amylopectin	Waxy corn	1700[b]	116[f]
Hydrolyzed amylopectin	Waxy corn	15[b]	25[f]
Glycogen	Sweet corn	190[b]	15[f]
Hydrolyzed glycogen	Sweet corn	20[b]	6.3[f]
Glycogen	Rabbit liver	390[b]	6.6[g]

[a] From Goring (1962).
[b] $\bar{M}_w$ by light scattering.
[c] $\bar{M}_n$ by viscometry from $[\eta] = 0.0091$ DP.
[d] Calculated from fractionation data.
[e] $\bar{M}_n$ by osmometry.
[f] $\bar{M}_n$ from the alkali number.
[g] $\bar{M}_{ww}$ from sedimentation and diffusion used instead of $\bar{M}_n$; usually $\bar{M}_w > \bar{M}_{ww} > \bar{M}_n$.

weight, is evidently polydisperse, being similar in this respect to the hemicelluloses.

3.2.3   The Configuration of Cellulose Molecules

Based on properties in solution such as intrinsic viscosity and sedimentation and diffusion rates, conclusions can be drawn concerning the polymer configuration. Like most of the synthetic polymers, such as polystyrene, cellulose in solution belongs to a group of linear, randomly coiling polymers. This means that the molecules have no preferred structure in solution in contrast to amylose and some protein molecules which can adopt helical conformations. Cellulose differs distinctly from synthetic polymers and from lignin in some of its polymer properties. Typical of its solutions are the comparatively high viscosities and low sedimentation and diffusion coefficients (Tables 3-2 and 3-3).

Any linear polymer molecule, even a reasonably stiff rod, will coil randomly, provided the chain is sufficiently long. In addition to the size of the monomer units, the tendency for coiling is affected by the forces between the units as well as the interaction between the polymer and the solvent. One measure of the stiffness of a polymer molecule is the end-to-end distance (R). For a polydisperse polymer the root-mean-square average of R ($\bar{R}^2)^{1/2}$ is used. R is affected by the properties of the polymer itself as well as by the interaction of the solvent. The better the solvent the more the polymer

TABLE 3-2. Examples of Intrinsic Viscosities Corresponding to a Molecular Weight $(\bar{M}_w)$ of 50,000 for Various Macromolecules[a]

Macromolecule	Solvent	Intrinsic viscosity $[\eta]$ (dm³/kg)
Dioxane-HCl lignin	Pyridine	8
Lignosulfonate	0.1 M NaCl	5
Kraft lignin	Dioxane	6
Alkali lignin	0.1 M buffer	4
Polymethylmethacrylate	Benzene	23
Polymethylstyrene	Toluene	24
Xylan	CED	216
Cellulose	CED	181

[a] From Goring (1971).

swells, whereas in a poor solvent contraction occurs (Fig. 3-11). The expansion tendency of a polymer molecule is characterized by Flory's equation $R = \alpha R_0$, where α is the expansion coefficient. At a certain temperature in a given solvent an "ideal" state ($R = R_0$) can be reached, where the environment has no observable influence on the polymer. Such a solvent is called a theta solvent and the temperature the theta or Flory temperature.

The intrinsic viscosity of a polymer such as cellulose is related to the molecular weight by the Mark–Houwink equation:

$$[\eta] = K\bar{M}_v{}^a$$

where the coefficient K and the exponent a are experimental constants characteristic for the solvent and polymer type and $\bar{M}_v$ the molecular weight. The value of exponent a may vary according to the configuration of the

TABLE 3-3. Typical Polymer Parameters of Cellulose and Hydroxyethyl Cellulose Compared with Polyvinyl Acetate[a]

Sample	Solvent	DP_w	$[\eta]$ (dm³/kg)	Sedimentation rate, s_0 (Svedbergs)	Diffusion rate, $D_0 \times 10^7$
Cellulose	Cadoxen	2290	645	3.8	0.58
Hydroxyethyl cellulose	Water	2180	895	4.9	0.96
Polyvinyl acetate	2-Butanone	4880	125	13.5	2.40

[a] From Brown (1966). © 1966. TAPPI. Reprinted from *Tappi* **49**(8), pp. 367–368, with permission.

 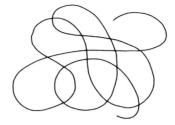

Fig. 3-11. Schematic representation of randomly coiling macromolecules in solution. In a good solvent (right) the interaction with the polymer is thermodynamically favorable, resulting in expansion whereas in a poor solvent the coil is rather compact (left) (Brown, 1966). © 1966. TAPPI. Reprinted from *Tappi* **49**(8), pp. 367–368, with permission.

polymer and is 0.6–0.8 for both cellulose and many synthetic polymers, such as polystyrene (cf. Table 3-4). For some cellulose derivatives like nitrates, a approaches the value 1.0.

At theta conditions where the molecule exists in the form of a completely closed-up (compact) random coil ($R = R_0$) a is 0.5. In a good solvent a is 0.8 and maximally about 1.0. For rods a is 1.8.

The intrinsic viscocity $[\eta]$ is a measure of the effective hydrodynamic volume of the molecule. For a hard, nonswelling sphere, Einstein's equation is valid:

$$[\eta] = 0.025\bar{V}$$

where $\bar{V}$ is the specific volume of the material in the sphere. In the case of linear, solvent-swollen polymers such as cellulose, $\bar{V}$ and thus also $[\eta]$ is

TABLE 3-4. Some Values for the Exponent a^a

Sample	Solvent	a
Dioxane-HCl lignin	Pyridine	0.15
Alkali lignin	Dioxane	0.12
Lignosulfonate	0.1 M NaCl	0.32
Polymethylstyrene	Toluene	0.71
Cellulose	Cadoxen	0.75
Cellulose nitrate	Ethyl acetate	0.99
Xylan	DMSO	0.94
Xylan	CED	1.15
Einstein sphere		0
Compact coil		0.5
Free-draining coil		1
Rods		1.8

^{a} From Goring (1971).

much larger. A low $[\eta]$ value means that the molecule is compact and thus occupies a relatively small volume. This is typical of lignin (see Section 4.4.2).

3.3 Hemicelluloses

Hemicelluloses were originally believed to be intermediates in the biosynthesis of cellulose. Today it is known, however, that hemicelluloses belong to a group of heterogeneous polysaccharides which are formed through biosynthetic routes different from that of cellulose. In contrast to cellulose which is a homopolysaccharide, hemicelluloses are heteropolysaccharides. Like cellulose most hemicelluloses function as supporting material in the cell walls. Hemicelluloses are relatively easily hydrolyzed by acids to their monomeric components consisting of D-glucose, D-mannose, D-xylose, L-arabinose, and small amounts of L-rhamnose in addition to D-glucuronic acid, 4-O-methyl-D-glucuronic acid, and D-galacturonic acid. Most hemicelluloses have a degree of polymerization of only 200.

Some wood polysaccharides are extensively branched and are readily soluble in water. Typical of certain tropical trees is a spontaneous formation of exudate gums, which are exuded as viscous fluids at sites of injury and after dehydration give hard, clear nodules rich in polysaccharides. These gums, for example, gum arabic, consist of highly branched, water-soluble polysaccharides.

The amount of hemicelluloses of the dry weight of wood is usually between 20 and 30% (cf. Appendix). The composition and structure of the hemicelluloses in the softwoods differ in a characteristic way from those in the hardwoods. Considerable differences also exist in the hemicellulose content and composition between the stem, branches, roots, and bark.

3.3.1 Softwood Hemicelluloses

Galactoglucomannans Galactoglucomannans are the principal hemicelluloses in softwoods (about 20%). Their backbone is a linear or possibly slightly branched chain built up by (1 → 4)-linked β-D-glucopyranose and β-D-mannopyranose units (Fig. 3-12). Galactoglucomannans can be roughtly divided into two fractions having different galactose contents. In the fraction which has a low galactose content the ratio galactose:glucose:mannose is about 0.1:1:4 whereas in the galactose-rich fraction the corresponding ratio is 1:1:3. The former fraction with a low galactose content is often referred to as glucomannan. The α-D-galactopyranose residue is linked as a single-unit side chain to the framework by (1 → 6)-bonds. An important

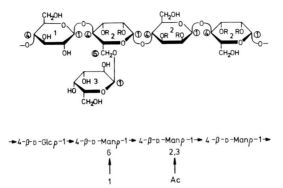

Fig. 3-12. The structure of galactoglucomannans. Sugar units: 1, β-D-glucopyranose (Glcp); 2, β-D-mannopyranose (Manp); 3, β-D-galactopyranose (Galp). R = CH₃CO or H. Below the abbreviated formula showing the proportions of the units (galactose-rich fraction).

structural feature is that the C-2 and C-3 positions in mannose and glucose units are partially substituted by acetyl groups, on the average one group per 3–4 hexose units. Galactoglucomannans are easily depolymerized by acids and especially so the bond between galactose and the main chain. The acetyl groups are much more easily cleaved by alkali than by acid.

Arabinoglucuronoxylan   In addition to galactoglucomannans, softwoods contain an arabinoglucuronoxylan (5–10%). It is composed of a framework containing (1 → 4)-linked β-D-xylopyranose units which are partially substituted at C-2 by 4-O-methyl-α-D-glucuronic acid groups, on the average two residues per ten xylose units. In addition, the framework contains α-L-arabinofuranose units, on the average 1.3 residues per ten xylose units (Fig. 3-13). Because of their furanosidic structure, the arabinose side chains are easily hydrolyzed by acids. Both the arabinose and uronic acid substituents stabilize the xylan chain against alkali-catalyzed degradation (see Sections 2.5.5 and 7.3.5).

Arabinogalactan   The heartwood of larches contains exceptionally large amounts of water-soluble arabinogalactan, which is only a minor constituent in other wood species. Its backbone is built up by (1 → 3)-linked β-D-galactopyranose units. Almost every unit carries a branch attached to position 6, largely (1 → 6)-linked β-D-galactopyranose residues but also L-arabinose (Fig. 3-14). There are also a few glucuronic acid residues present in the molecule. The highly branched structure is responsible for the low viscosity and high solubility in water of this polysaccharide.

Other Polysaccharides   Besides galactoglucomannans, arabinoglucuronoxylan and arabinogalactan, softwoods contain other polysaccharides, usually present in minor quantities. They are built up dominantly

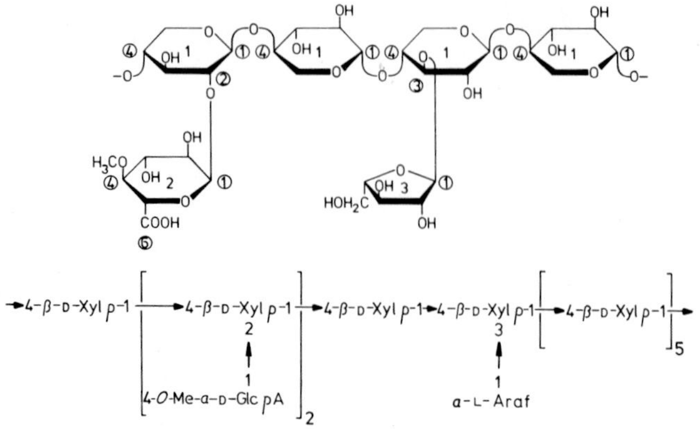

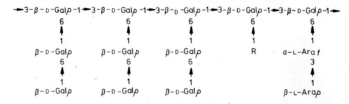

Fig. 3-13. The structure of arabinoglucuronoxylan. Sugar units: 1, β-D-xylopyranose (Xylp); 2, 4-O-methyl-α-D-glucopyranosyluronic acid (GlcpA); 3, α-L-arabinofuranose (Araf). Below are the abbreviated formulas showing the proportions of the units.

by units of arabinose, galactose, glucuronic, and galacturonic acids and include starch and pectins. In addition, compression wood contains about 10% (1 → 4)-β-D-galactan and 2–5% (1 → 3)-β-D-glucan (laricinan).

3.3.2 Hardwood Hemicelluloses

Glucuronoxylan Even if hemicelluloses in various hardwood species differ from each other both quantitatively and qualitatively, the major component is an *O*-acetyl-4-*O*-methylglucurono-β-D-xylan, sometimes called glucuronoxylan. Often the xylose-based hemicelluloses in both softwoods and hardwoods are termed simply xylans.

Depending on the hardwood species, the xylan content varies within the limits of 15–30% of the dry wood. As can be seen from Fig. 3-15, the backbone consists of β-D-xylopyranose units, linked by (1 → 4)-bonds. Most of the xylose residues contain an acetyl group at C-2 or C-3 (about seven

Fig. 3-14. Abbreviated formula of arabinogalactan. Sugar units: β-D-galactopyranose (Galp), β-L-arabinopyranose (Arap), α-L-arabinofuranose (Araf), and R is β-D-galactopyranose or, less frequently, α-L-arabinofuranose, or a β-D-glucopyranosyluronic acid residue.

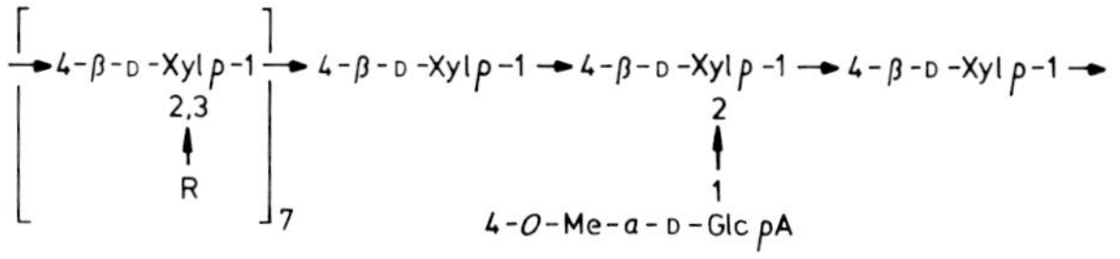

Fig. 3-15. Abbreviated formula of glucuronoxylan. Sugar units: β-D-xylopyranose (Xyl$p$), and 4-O-methyl-α-D-glucopyranosyluronic acid (Glc$p$A). R is an acetyl group (CH$_3$CO).

acetyl residues per ten xylose units). The xylose units in the xylan chain additionally carry (1 → 2)-linked 4-O-methyl-α-D-glucuronic acid residues, on the average about one uronic acid per ten xylose residues. The xylosidic bonds between the xylose units are easily hydrolyzed by acids, whereas the linkages between the uronic acid groups and xylose are very resistant. Acetyl groups are easily cleaved by alkali, and the acetate formed during kraft pulping of wood mainly originates from these groups (see Section 7.3.5). They are slowly hydrolyzed to acetic acid within a living tree as a result of the acidic nature of especially the heartwood (cf. also galactoglucomannans, p. 60).

Recent studies have revealed certain interesting features in the structure of a birch xylan (Fig. 3-16). The unit next to the reducing xylose end group is D-galacturonic acid, linked to a L-rhamnose unit through the C-2 position. The rhamnose unit, in turn, is connected through its C-3 position to the xylan chain.

Glucomannan Besides xylan, hardwoods contain 2–5% of a glucomannan, which is composed of β-D-glucopyranose and β-D-mannopyranose units linked by (1 → 4)-bonds (Fig. 3-17). The glucose:mannose ratio varies between 1:2 and 1:1, depending on the wood species. The mannosidic bonds between the mannose units are more rapidly hydrolyzed by acid than the corresponding glucosidic bonds, and glucomannan is easily de-polymerized under acidic conditions.

Other Polysaccharides In addition to xylan and glucomannan minor amounts of miscellaneous polysaccharides are present in hardwoods, partly of the same type as those occurring in softwoods. They might be important components for the living tree, although of little interest when considering the technical applications.

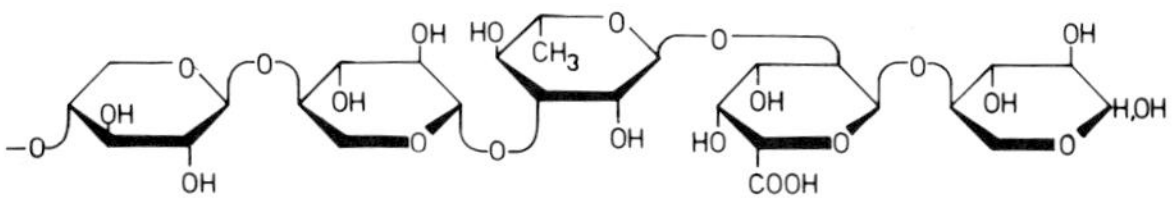

$-\beta$-D-Xyl p-1$\rightarrow$4-β-D-Xyl p-1$\rightarrow$3-α-L-Rha p-1$\rightarrow$2-α-D-Gal pA-1$\rightarrow$4-D-Xyl p

Fig. 3-16. Structure associated with the reducing end group of birch xylan (Johansson and Samuelson, 1977).

→4-β-D-Glc p-1→4-β-D-Man p-1→4-β-D-Glc p-1→4-β-D-Man p-1→4-β-D-Man p-1→

Fig. 3-17.   Abbreviated formula of glucomannan.

3.3.3   Summary of Wood Hemicelluloses

Table 3-5 summarizes the main structural features of the hemicelluloses appearing in both softwoods and hardwoods.

3.3.4   Isolation of Hemicelluloses

Hemicelluloses can be isolated from wood, holocellulose, or pulp by extraction. Among the few neutral solvents which are effective, dimethyl sulfoxide is useful particularly for the extraction of xylan from a holocellulose. Although only a part of the xylan can be extracted, the advantage is that no chemical changes take place. More xylan can be extracted with alkali (KOH or NaOH). Addition of sodium borate to the alkali facilitates the dissolution of galactoglucomannans and glucomannans. However, alkali extractions have the disadvantage of deacetylating the hemicelluloses completely.

Glucomannans are more effectively extracted by sodium than by potassium hydroxide. A gradient elution at varying alkali concentrations can be used for a rough fractionation of the hemicellulose components. The solvating effect of borate ions is based on their reaction with vicinal hydroxyl groups in the *cis* position present, for example, in mannose units. The borate complex is readily decomposed on acidification. The polysaccharides can be precipitated from the alkaline extract by acidification with acetic acid. Addition of a neutral organic solvent, e.g., ethyl alcohol, to the neutralized extraction solution results in a more complete precipitation. Some more specific precipitation agents are also known, for instance, barium hydroxide for glucomannan and cetyltrimethylammonium bromide or hydroxide for glucuronoxylan. Fehling's solution can also be used for precipitation purposes.

The precipitated hemicellulose preparations can be further purified by column chromatography. Gel permeation chromatography is useful for fractionation according to the molecular weight. In most cases, however, differences in chemical properties form the basis for the separation, for example, the above-mentioned capability of certain hydroxyl groups for complex formation. Particularly useful in the separation of hemicelluloses are ion exchangers based on cellulose, dextran or agarose, e.g., diethylaminoethyl cellulose in different ionic forms. Generally, chromatography in its various forms is used for the characterization of the acidic hydrolysis products of isolated hemicelluloses, either after total hydrolysis (analysis of

TABLE 3-5. The Major Hemicellulose Components

Hemicellulose type	Occurrence	Amount (% of wood)	Composition			Solubility[a]	$\overline{DP}_n$
			Units	Molar ratios	Linkage		
Galactoglucomannan	Softwood	5–8	β-D-Manp	3	1 → 4	Alkali, water*	100
			β-D-Glcp	1	1 → 4		
			α-D-Galp	1	1 → 6		
			Acetyl	1			
(Galacto)glucomannan	Softwood	10–15	β-D-Manp	4	1 → 4	Alkaline borate	100
			β-D-Glcp	1	1 → 4		
			α-D-Galp	0.1	1 → 6		
			Acetyl	1			
Arabinoglucuronoxylan	Softwood	7–10	β-D-Xylp	10	1 → 4	Alkali, dimethylsulfoxide*, water*	100
			4-O-Me-α-D-ClcpA	2	1 → 2		
			α-L-Araf	1.3	1 → 3		
Arabinogalactan	Larch wood	5–35	β-D-Galp	6	1 → 3, 1 → 6	Water	200
			α-L-Araf	2/3	1 → 6		
			β-D-Arap	1/3	1 → 3		
			β-D-GlcpA	Little	1 → 6		
Glucuronoxylan	Hardwood	15–30	β-D-Xylp	10	1 → 4	Alkali, dimethylsulfoxide*	200
			4-O-Me-α-D-GlcpA	1	1 → 2		
			Acetyl	7			
Glucomannan	Hardwood	2–5	β-D-Manp	1–2	1 → 4	Alkaline borate	200
			β-D-Glcp	1	1 → 4		

[a] The asterisk represents a partial solubility.

monosaccharides) or after partial hydrolysis (analysis of oligosaccharides). The general method for the localization of bonds is complete methylation followed by hydrolysis and identification of the methylated sugars (gas-liquid chromatography–mass spectrometry). In addition, separate determinations can be carried out for the analysis of uronic acids, pentosan, acetyl, and methoxyl groups as well as for the molecular weight and its distribution. (see also Section 2.6.1.)

3.3.5   Distribution of Hemicelluloses in Wood

By determining the polysaccharide composition of cells microtomed from the cambial layer, representing different stages of growth, it has been possible to determine approximately the distribution of polysaccharides in different layers of the cell walls. In softwoods (spruce and pine), the xylan content seems to be lowest in the middle layer of the secondary wall (S_2) and considerably higher in the S_1 and S_3 layers. The opposite is true for the distribution of galactoglucomannans. In hardwood (birch), the xylan content is highest in the S_2 layer. The hemicellulose composition is different in tracheids and fibers on the one hand and ray cells on the other. In both hardwoods and softwoods, the ray cells have a higher proportion of xylan than do the tracheids and fibers. In spruce and pine, the ray cells even contain more xylan than glucomannan. The tension wood of some hardwoods contains a complex galactan. It is found in birch and beech but not in maple or poplars (aspen). Finally, the hemicellulose composition and content vary between young cells which are still growing and completely developed mature cells. Juvenile wood contains more xylan and less cellulose and glucomannan than mature wood. In the softwoods earlywood contains more xylan and less glucomannan than latewood.

Probably no chemical bonds exist between cellulose and hemicelluloses but sufficient mutual adhesion is provided by hydrogen bonds and van der Waals forces. Chemical bonds obviously exist between hemicelluloses and lignin (see Section 4.2.7).

References

Aspinall, G. O. (1970). Pectins, plants gums and other plant polysaccharides. In "The Carbohydrates" (W. Pigman and D. Horton, eds.), 2nd ed., Vol. 2B, pp. 515–536. Academic Press, New York.

Bikales, N. M., and Segal, L., eds. (1971). "Cellulose and Cellulose Derivatives, Parts IV–V," 2nd ed. Wiley (Interscience), New York.

Billmeyer, F. W., Jr. (1965). Characterization of molecular weight distributions in high polymers. J. Polym. Sci., Part C, No. 8, 161–178.

Blackwell, J., Kolpak, F. J., and Gardner, K. H. (1977). Structures of native and regenerated celluloses. In "Cellulose Chemistry and Technology" (J. C. Arthur, Jr., ed.), ACS Symposium Series, No. 48, pp. 42–55. Am. Chem. Soc., Washington, D.C.

Brown, W. J. (1966). The configuration of cellulose and derivatives in solution. *Tappi* **49,** 367–373.

Browning, G. L. (1967). "Methods of Wood Chemistry," Vol. 2, pp. 561–587. Wiley (Interscience), New York.

Delmer, D. P. (1980). Cellulose synthesis. *In* "CRS Handbook Series of Biosolar Resources, Vol. 1, Basic Principles" (C. C. Black, A. Mitsui, and O. R. Zaborsky, eds.). CRS Press, Boca Raton, Florida (in press).

Gardner, K. H., and Blackwell, J. (1974). The hydrogen bonding in native cellulose. *Biochim. Biophys. Acta* **343,** 232–237.

Goring, D. A. I. (1962). The physical chemistry of lignin. *Pure Appl. Chem.* **5,** 233–254.

Goring, D. A. I. (1971). Polymer properties of lignin and lignin derivatives. *In* "Lignins (K. V. Sarkanen and C. H. Ludwig, eds.), pp. 695–768. Wiley (Interscience), New York.

Hassid, W. Z. (1970). Biosynthesis of sugars and polysaccharides. *In* "The Carbohydrates" (W. Pigman and D. Horton, eds.), 2nd ed., Vol. 2A, pp. 301–373. Academic Press, New York.

Johansson, M. H., and Samuelson, O. (1977). Reducing end groups in birch xylan and their alkaline degradation. *Wood Sci. Technol.* **11,** 251–263.

Kolpak, F. J., and Blackwell, J. (1976). Determination of the structure of cellulose II. *Macromolecules* **9,** 273–278.

Kolpak, F. J., Weih, M., and Blackwell, J. (1978). Mercerization of cellulose: 1. Determination of the structure of mercerized cotton. *Polymer* **19,** 123–131.

Leloir, L. F. (1964). The biosynthesis of polysaccharides. *Proc. Plenary Sess. Int. Congr. Biochem., 6th, New York*, pp. 15–29.

Meier, H. (1962). Studies on a galactan from tension wood of beech (*Fagus Silvatica L.*). *Acta Chem. Scand.* **16,** 2275–2283.

Meier, H., and Wilkie, K. C. B. (1959). The distribution of polysaccharides in the cell-wall of tracheids of pine (*Pinus Silvestris L.*). *Holzforschung* **13,** 177–182.

Miller, L. P., ed. (1973). "Phytochemistry," Vol. 1. Van Nostrand-Reinhold, New York.

Nikaido, H., and Hassid, W. Z. (1971). Biosynthesis of saccharides from glucopyranosyl esters of nucleoside pyrophosphates ("sugar nucleotides"). *Adv. Carbohydr. Chem. Biochem.* **26,** 351–483.

Ott, E., Spurlin, H. M., and Grafflin, M. W., eds. (1954). "Cellulose and Cellulose Derivatives, Parts I–III," 2nd ed. Wiley (Interscience), New York.

Overend, W. G. (1972). Gycosides. *In* "The Carbohydrates" (W. Pigman and D. Horton, eds.), 2nd ed., Vol. 1A, pp. 279–353. Academic Press, New York.

Sjöström, E., and Enström, B. (1967). Characterization of acidic polysaccharides isolated from different pulps. *Tappi* **50,** 32–36.

Sumi, Y., Hale, R. D., and Rånby, B. G. (1963). The accessibility of native cellulose microfibrils. *Tappi* **46,** 126–130.

Sumi, Y., Hale, R. D., Meyer, J. A., Leopold, B., and Rånby, B. G. (1964). Accessibility of wood and wood carbohydrates measured with tritiated water. *Tappi* **47,** 621–624.

Timell, T. E. (1964). Wood hemicelluloses, Part I. *Adv. Carbohydr. Chem. Biochem.* **19,** 247–302.

Timell, T. E. (1965). Wood hemicelluloses, Part II. *Adv. Carbohydr. Chem. Biochem.* **20,** 409–483.

Timell, T. E. (1965). Wood and bark polysaccharides. *In* "Cellular Ultrastructure of Woody Plants" (W. A. Côté, Jr., ed.), pp. 127–156. Syracuse Univ. Press, Syracuse, New York.

Timell, T. E. (1967). Recent progress in the chemistry of wood hemicelluloses. *Wood Sci. Technol.* **1,** 45–70.

Whistler, R. L., and Richards, E. L. (1970). Hemicelluloses. *In* "The Carbohydrates" (W. Pigman and D. Horton, eds.), 2nd ed., Vol. 2A, pp. 447–469. Academic Press, New York.

Chapter 4

LIGNIN

Anselme Payen observed in 1838 that wood, when treated with concentrated nitric acid, lost a portion of its substance, leaving a solid and fibrous residue he named cellulose. As a result of much later studies it became evident that the fibrous material isolated by Payen contained also other polysaccharides besides cellulose. The dissolved material (*"la matière incrustante"*), on the other hand, had a higher carbon content than the fibrous residue and was termed "lignin" in 1865 by F. Schulze who derived the name from the latin word for wood (lignum).

Later, the development of technical pulping processes generated much interest in lignin and its reactions. In 1897, Peter Klason studied the composition of lignosulfonates and put forward the idea that lignin was chemically related to coniferyl alcohol. In 1907, he proposed that lignin is a macromolecular substance and, ten years later, that coniferyl alcohol units are joined together by ether linkages.

4.1   Isolation of Lignin

Lignin can be isolated from extractive-free wood as an insoluble residue after hydrolytic removal of the polysaccharides. Alternatively, lignin can be hydrolyzed and extracted from the wood or converted to a soluble derivative. So-called Klason lignin is obtained after removing the polysaccharides from extracted (resin-free) wood by hydrolysis with 72% sulfuric acid. Other acids can be used as well for the hydrolysis, but the method has the serious drawback in that the structure of lignin is extensively changed during the hydrolysis. The polysaccharides may also be removed by enzymes from finely divided wood meal. The method is tedious, but the resulting "cellulolytic enzyme lignin" (CEL) retains its original structure essentially unchanged. Lignin can also be extracted from wood using dioxane containing water and hydrochloric acid, but considerable changes in its structure occur.

Besides cellulolytic enzyme lignin, the so-called Björkman lignin, alternatively referred to as "milled wood lignin" (MWL) is the best preparation known so far, and it has been widely used for structural studies. When wood meal is ground in a ball mill either dry or in the presence of nonswelling solvents, e.g., toluene, the cell structure of the wood is destroyed and a portion of lignin (usually not more than 50%) can be obtained from the suspension by extraction with a dioxane–water mixture. MWL preparations always contain some carbohydrate material.

Soluble lignin derivatives (lignosulfonates) are formed by treating wood at elevated temperatures with solutions containing sulfur dioxide and hydrogen sulfite ions (see Section 7.2). Lignin is also dissolved as alkali lignin when wood is treated at elevated temperatures (170°C) with sodium hydroxide, or better, with a mixture of sodium hydroxide and sodium sulfide (sulfate or kraft lignin) (see Section 7.3). Lignin is further converted to an alkali-soluble derivative by a solution of hydrochloric acid and thioglycolic acid at 100°C.

Softwood lignin can be determined gravimetrically by the Klason method. Normal softwood contains 26–32% lignin while the lignin content of compression wood is 35–40% (cf. Appendix). The lignin present in hardwoods is partly dissolved during the acid hydrolysis and hence the gravimetric values must be corrected for the "acid-soluble lignin" using UV spectrophotometry. Direct UV spectrophotometric methods have also been developed for the determination of lignin in wood and pulps. Normal hardwood contains 20–28% lignin, although tropical hardwoods can have a lignin content exceeding 30%. Tension wood contains only 20–25% lignin.

4.2   The Formation and Structure of Lignin

Lignins are polymers of phenylpropane units. Many aspects in the chemistry of lignin still remain unclear, for example, the specific structural features of lignins located in various morphological regions of the woody xylem. Nevertheless, the principal structural elements in lignins have been largely clarified as a result of detailed studies on isolated lignin preparations, such as milled wood lignin, using specific degradative techniques based on oxidation, reduction, or hydrolysis under acidic and alkaline conditions. Much effort has been directed toward the clarification of the biosynthesis of lignin. Detailed identification of the reaction products has been possible by novel chromatographic techniques and spectroscopic methods developed during the last two decades.

4.2.1 Phenylpropane—The Basic Structural Unit of Lignin

Methods based on classical organic chemistry led to the conclusion, already by 1940, that lignin is built up of phenylpropane units. Examples of typical reactions used in these studies are illustrated in Fig. 4-1. However, the concept of a phenylpropanoid structure failed to win unanimous acceptance, and as late as 30 years ago, some scientists were not convinced that lignin in its native state was an aromatic substance. Finally, the problem was solved by Lange in 1954, who applied UV microscopy at various wavelengths directly on thin wood sections, obtaining spectra typical of aromatic compounds.

4.2.2 The Biosynthesis of Lignin Precursors

The role of coniferyl alcohol as the immediate precursor of softwood lignin has been demonstrated by using ^{14}C labeling. Administration of labeled coniferyl alcohol as β-glucoside (coniferin) to seedlings of spruce results in the exclusive formation of radioactive lignin.

Lignin precursors (p-coumaryl, coniferyl, and sinapyl alcohols) are formed from glucose by a variety of enzymic reactions involving oxidations, reductions, aminations, deaminations, decarboxylations, etc. (Fig. 4-2). D-Glucose generated in photosynthesis is transformed first to a heptose phosphate derivative which then cyclizes to 5-dehydroquinic acid. The reaction sequence leads ultimately to phenylalanine with shikimic and phenylpyruvic acids as intermediates. This series of reactions is known as the shikimic acid route. In the Gramineae including grasses, on the other hand, lignin is also formed through an alternative pathway proceeding via tyrosine (p-hydroxyphenylalanine). Phenylalanine is deaminated to cinnamic acid which then acquires aromatic hydroxyl and methoxyl groups. The final precursors are formed after reduction of the carboxyl group to a primary alcohol. The precursors are probably present in the cambial tissues of gymnosperms as glucosides and become liberated by the action of a β-glucosidase.

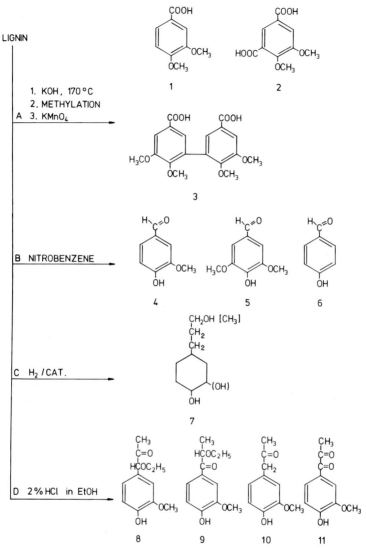

Fig. 4-1. Examples of classical methods indicating a phenylpropanoid structure of lignin. (A) *Permanganate oxidation* (methylated spruce lignin) affords veratric acid (3,4-dimethoxybenzoic acid) (1) in a yield of 8% and minor amounts of isohemipinic (4,5-dimethoxyisophtalic acid) (2) and dehydrodiveratric (3) acids. The formation of isohemipinic acid supports the occurrence of condensed structures (e.g., β-5 or γ-5). (B) *Nitrobenzene oxidation* of softwoods in alkali results in the formation of vanillin (4-hydroxy-3-methoxybenzaldehyde) (4) (about 25% of lignin). Oxidation of hardwoods and grasses results, respectively, in syringaldehyde (3,5-dimethoxy-4-hydroxybenzaldehyde) (5) and p-hydroxybenzaldehyde (6). (C) *Hydrogenolysis* yields propylcyclohexane derivatives (7). (D) *Ethanolysis* yields so-called Hibbert ketones (8,9,10, and 11).

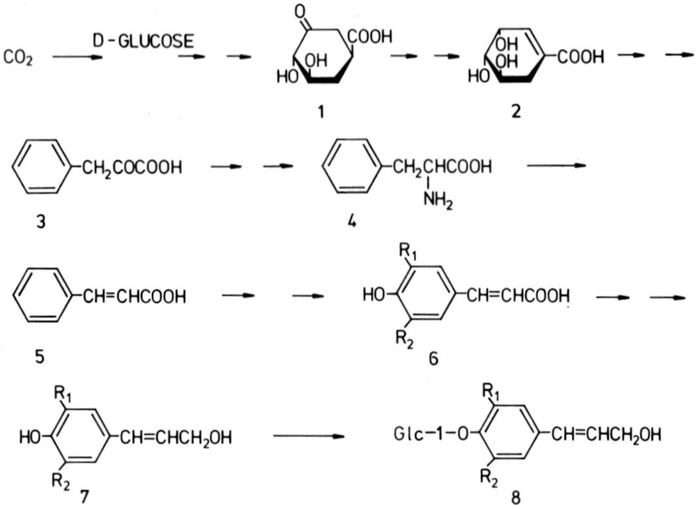

Fig. 4-2. Simplified reaction route illustrating the formation of lignin precursors. 1, 5-Dehydroquinic acid; 2, shikimic acid; 3, phenylpyruvic acid; 4, phenylalanine; 5, cinnamic acid; 6, ferulic acid (R_1=H and R_2=OCH$_3$), sinapic acid (R_1=R_2=OCH$_3$), and p-coumaric acid (R_1=R_2=H); 7, coniferyl alcohol (R_1=H and R_2=OCH$_3$), sinapyl alcohol (R_1=R_2=OCH$_3$), and p-coumaryl alcohol (R_1=R_2=H); 8, the corresponding glucosides of 7.

4.2.3 Biosynthesis of Lignin

For an understanding of the formation and structure of lignin, investigations conducted by H. Erdtman in 1930 were of great importance. He studied the oxidative dimerization of various phenols in the biogenesis of natural products and reached the conclusion that lignin must be formed from α,β-unsaturated C_6C_3 precursors of the coniferyl alcohol type (Fig. 4-3) via enzymic dehydrogenation. The polymerization of precursors to lignin in nature does indeed occur in this manner, as has been demonstrated by comprehensive studies by Freudenberg and co-workers during the period 1940 to 1970.

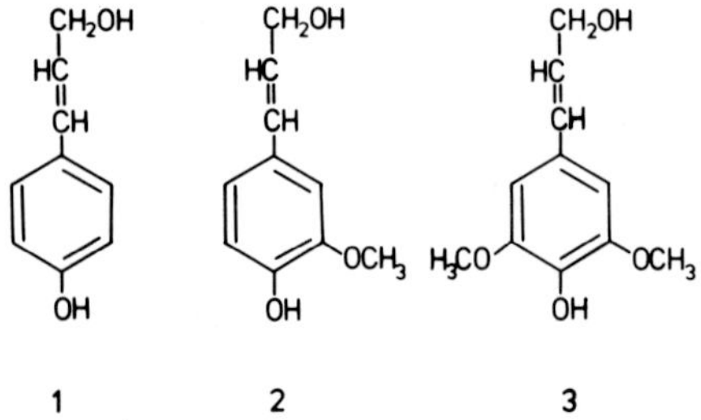

Fig. 4-3. Lignin precursors. 1, p-Coumaryl alcohol, 2, coniferyl alcohol; 3, sinapyl alcohol.

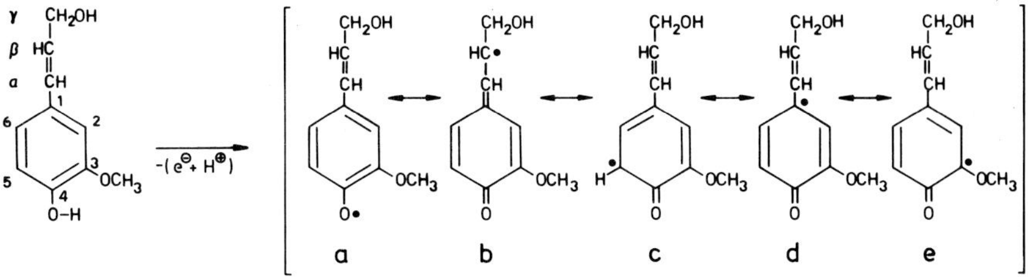

Fig. 4-4.   Formation of resonance-stabilized phenoxy radicals by the enzymic dehydrogenation of coniferyl alcohol (Adler, 1977).

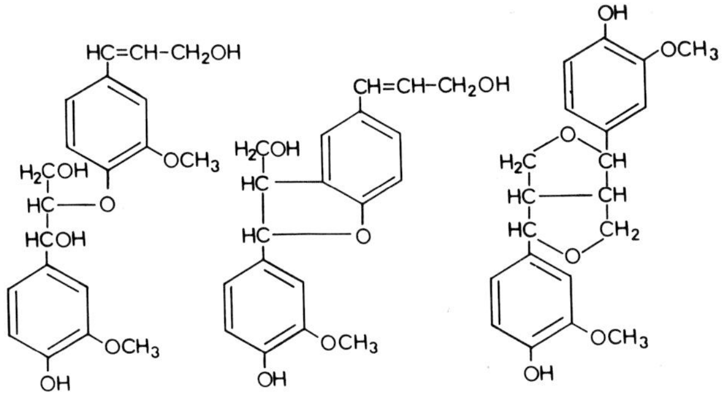

Fig. 4-5.   Formation of lignols from phenoxy radicals (cf. Fig. 4-4) (Adler, 1977).

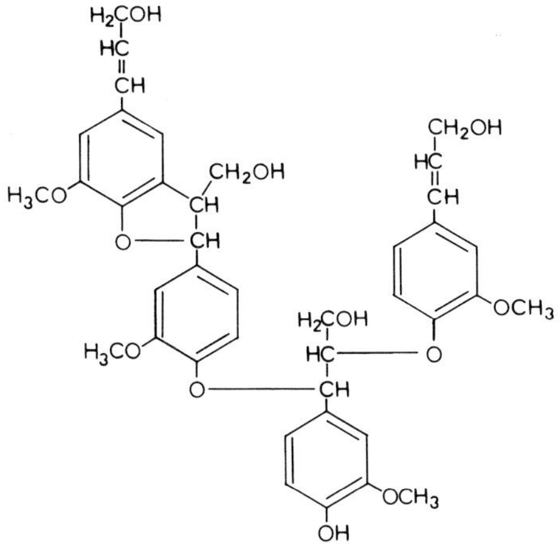

The enzymic dehydrogenation reaction is initiated by an electron transfer which results in the formation of resonance-stabilized phenoxy radicals (Fig. 4-4). The combination of these radicals produces a variety of dimers and oligomers, termed lignols (Fig. 4-5).

It can be readily shown that further oxidative coupling of di- and oligolignols ("*bulk* polymerization") would lead to a product containing a large number of unsaturated side chains. Since their amount in lignin is relatively low, the reaction presumably proceeds, after a certain initial period, essen-

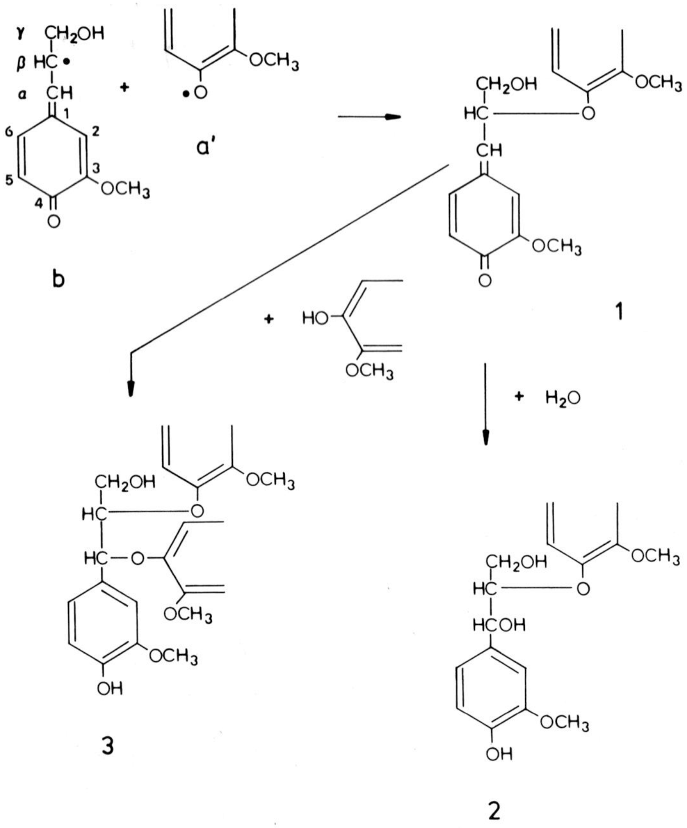

Fig. 4-6. Endwise polymerization (Adler, 1977). The monomeric radical b is attached by β-O-4 coupling to an end group radical a'. A guaiacylglyserol-β-aryl ether structure (2) is obtained from the intermediate (quinone methide) (1) by addition of water. As an alternative a phenolic hydroxyl group can be added to quinonemethide (1) resulting in a guaiacylglycerol-α,β-diaryl ether structure (3).

tially as "*endwise* polymerization." This means that the monomeric precursors are joined to the ends of the growing polymer instead of combining with each other. The endwise polymerization is also likely since the monomer concentration is probably quite low in the reaction zone.

Combination of the monomeric radicals to the phenolic end groups exclusively through β-O-4 and β-5 coupling modes would lead to a linear polymer (Figs. 4-6 and 4-7). However, branching of the polymer may take place through the formation of benzyl aryl ether structures (Figs. 4-5 and 4-6). In addition, 5-5 coupling to biphenyl structures and 5-O-4 coupling to diaryl ether units produce additional branched elements. The formation of biphenyl and diaryl ether structures primarily occurs in the coupling of two end group radicals rather than in the combination of monomer radicals to the end group radicals.

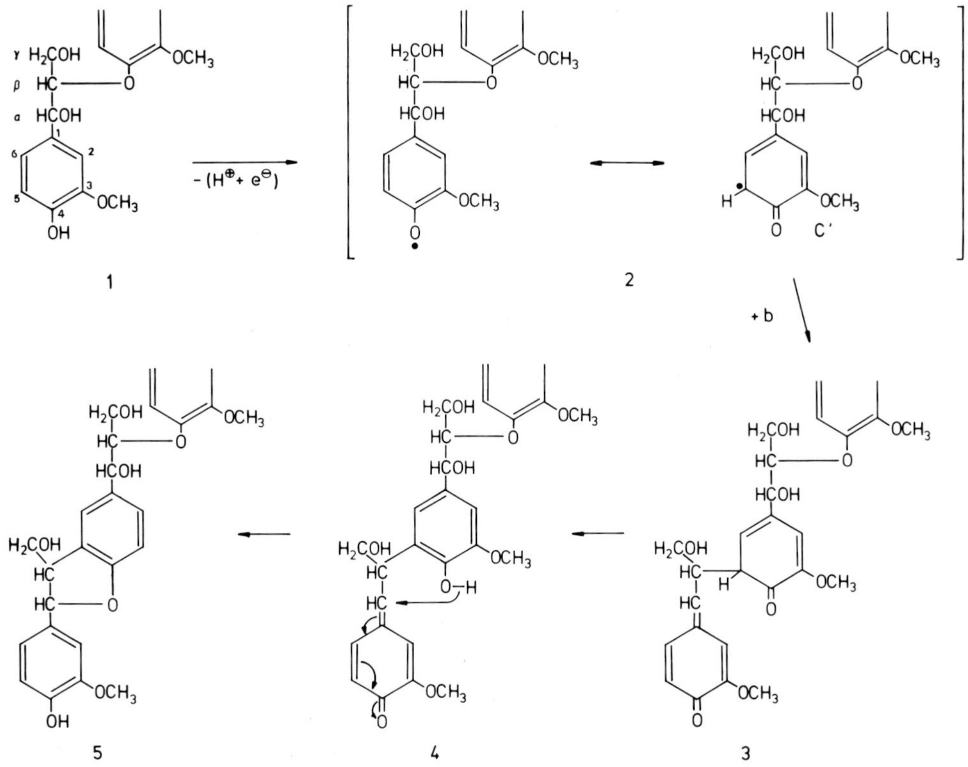

Fig. 4-7. Endwise polymerization (Adler, 1977). A guaiacylglycerol-β-aryl ether structure (1) is dehydrogenated and after resonance, radical c' is coupled with a coniferyl alcohol radical b (cf. Fig. 4-4). The β-5 coupling product (3) is tautomerized and undergoes intramolecular ring closure (a phenylcoumaran structure, 5).

4.2.4 Types of Linkages and Dimeric Structures

The dominating structures in the lignin molecule together with a variety of minor structural elements have been elucidated gradually as the methods for the identification of degradation products and for the synthesis of model compounds have been improved. Acid hydrolysis ("acidolysis") has proved to be more useful than ethanolysis (cf. Fig. 4-1D). A variety of dimers and oligomers have been identified among the acidolysis products of MWL, revealing the high frequency of such elements as the guaiacylglycerol-β-aryl ether and the phenylcoumaran structures. A particularly useful structural method consists of the oxidation of methylated lignin by permanganate and hydrogen peroxide to aromatic acids followed by identification of the reaction products by gas-liquid chromatography (cf. Fig. 4-1A).

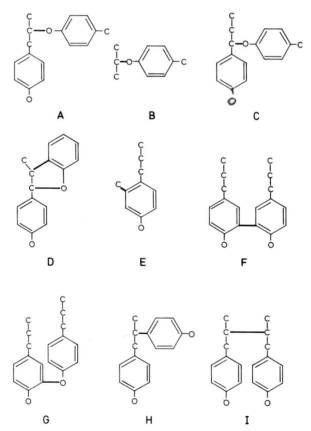

Fig. 4-8. The most common linkages between phenylpropane units. For proportions, see Tables 4-1 and 4-2.

TABLE 4-1.   Percentages of Different Types of Bonds in Spruce
(*Picea abies*) Lignin (MWL)[a]

	Bond type[b]	Percentage
A	Arylglycerol-β-aryl ether	48
B	Glyceraldehyde-2-aryl ether	2
C	Noncyclic benzyl aryl ether	6-8
D	Phenylcoumaran	9-12
E	Structures condensed in 2- or 6-positions	2.5-3
F	Biphenyl	9.5-11
G	Diaryl ether	3.5-4
H	1,2-Diarylpropane	7
I	β,β-linked structures	2

[a] From Adler (1977).
[b] Letters A through I are defined in Fig. 4-8.

More than two thirds of the phenylpropane units in lignin are linked by ether bonds, the rest by carbon-to-carbon bonds. Figure 4-8 shows the principal bonds in lignin. Their proportions are seen from Table 4-1 (spruce lignin) and Table 4-2 (birch lignin).

4.2.5   Functional Groups

The functional groups of major influence upon the reactivity of lignin consist of phenolic hydroxyl, benzylic hydroxyl, and carbonyl groups. Their frequency might vary according to the morphological location of the lignins. Some typical values are collected in Table 4-3.

TABLE 4-2.   Percentages of Different Types of Bonds in Birch
(*Betula verrucosa*) Lignin (MWL)[a]

Bond type[b]	Guaiacyl	Syringyl	Total
A	22-28	34-39	60
B			2
C			6-8
D			6
E	1-1.5	0.5-1	1.5-2.5
F	4.5		4.5
G	1	5.5	6.5
H			7
I			3

[a] From Adler (1977).
[b] Letters A through I are defined in Fig. 4-8.

TABLE 4-3.   Functional Groups of Lignin per 100 C_6C_3 Units

Group[a]	Spruce lignin	Birch lignin
Methoxyl	92-96	139-158
Phenolic hydroxyl (free)	15-30	9-13
Benzyl alcohol	15-20	
Noncyclic benzyl ether	7-9	
Carbonyl	20	

[a] The contents may vary depending on the origin of the lignin (e.g., middle lamella or secondary wall lignin).

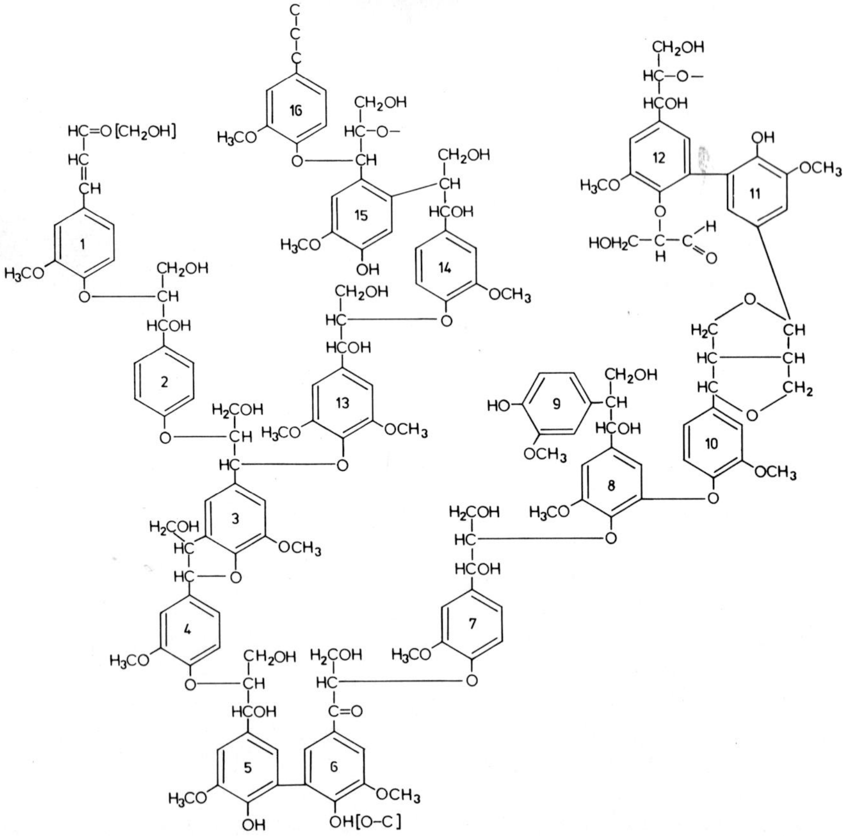

Fig. 4-9.   Prominent structures of softwood lignin (Adler, 1977) comprising 16 phenylpropane units.

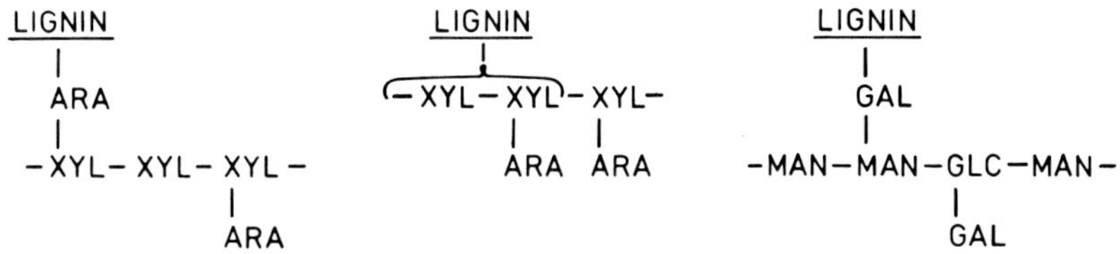

```
LIGNIN              LIGNIN                 LIGNIN
  |                   |                      |
 ARA              ⌐XYL─XYL⌐─XYL─            GAL
  |                  |    |                  |
-XYL─XYL─XYL─        ARA  ARA        -MAN─MAN─GLC─MAN─
     |                                       |
    ARA                                     GAL
```

Fig. 4-10. Examples of linkages between lignin and hemicelluloses (Eriksson and Lindgren, 1977). The uronic acid units in the xylan are not depicted.

4.2.6 The Structure of Lignin

Both Adler and Freudenberg have constructed structural schemes for spruce lignin and Nimz for beech lignin in which the main units and linkage types are represented in their approximate proportions. The scheme by Adler is depicted in Fig. 4-9.

4.2.7 Linkages between Lignin and Carbohydrates

Many studies have indicated that covalent linkages must exist between lignin and wood polysaccharides. Separation and analysis of lignin–carbohydrate complexes (LCC) have led to the conclusion that the hemicellulose components (xylan and galactoglucomannans in softwood) are bound to lignin mainly through arabinose, xylose, and galactose moieties as shown in Fig. 4-10.

4.3 Classification and Distribution of Lignin

Lignins can be divided into several classes according to their structural elements. So-called "guaiacyl lignin" which occurs in almost all softwoods is largely a polymerization product of coniferyl alcohol. The "guaiacyl-syringyl lignin," typical of hardwoods, is a copolymer of coniferyl and sinapyl alcohols, the ratio varying from 4:1 to 1:2 for the two monomeric units. An additional example is compression wood, which has a high proportion of phenylpropane units of the p-hydroxyphenyl type in addition to the normal guaiacyl units. The terms "syringyl lignin" and "p-hydroxyphenyl lignin" are sometimes used to denote the respective structural elements even if probably no natural lignins are exclusively composed of these units.

The lignin concentration is high in the middle lamella and low in the secondary wall. Because of its thickness, at least 70% of the lignin in softwoods is, however, located in the secondary wall as shown by quantitative UV microscopy (Fig. 4-11 and Table 4-4). The picture is very similar for

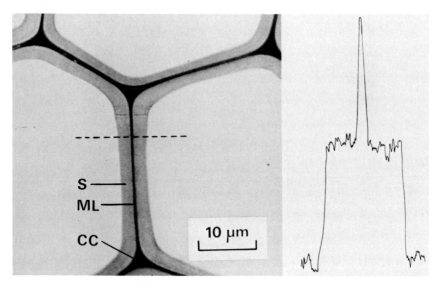

Fig. 4-11. Transverse section of a spruce tracheid photographed in UV light (240 nm) (Fergus *et al.*, 1969). The densitometer tracing has been taken across the tracheid wall along the dotted line. S, secondary wall; ML, compound middle lamella; CC cell corner.

the hardwoods (Table 4-5) although in this case analytical uncertainties are involved because of the more heterogeneous nature of the wood and the presence of both guaiacyl and syringyl units in the lignin. The measurements so far indicate that the lignin located in the secondary wall of hardwood fibers has a high content of syringyl units whereas larger amounts of guaiacyl units are present in the middle lamella lignin. The vessels in birch seem to contain only guaiacyl lignin, whereas syringyl lignin predominates in parenchyma cells.

TABLE 4-4. Distribution of Lignin in Spruce (Black Spruce, *Picea mariana*) Tracheid[a]

Wood	Morphological region[b]	Tissue volume (%)	Lignin (% of total)	Lignin concentration (%)
Earlywood	S	87	72	23
	ML	9	16	50
	CC	4	12	85
Latewood	S	94	82	22
	ML	4	10	60
	CC	2	9	100

[a] From Fergus *et al.* (1969).
[b] For explanations, see Fig. 4-11.

TABLE 4-5. Distribution of Lignin in Birch (White Birch, *Betula papyrifera*) Xylem[a]

Cell	Morphological region[b]	Type of lignin[c]	Tissue volume (%)	Lignin (% of total)	Lignin concentration (%)
Fiber	S	Sy	73	60	19
	ML	SyGu	5	9	40
	CC	SyGu	2	9	85
Vessel	S	Gu	8	9	27
	ML	Gu	1	2	42
Ray cell	CC	Sy	11	11	27

[a] From Fergus and Goring (1970).
[b] For explanations, see Fig. 4-11.
[c] Sy, syringyl lignin; SyGu, syringyl–guaiacyl lignin; Gu, guaiacyl lignin.

4.4 The Polymer Properties of Lignin and Its Derivatives

Most of the research on the macromolecular chemistry of lignin has been concentrated on lignosulfonates and kraft lignin because of the insolubility of lignin in its native state. The polymer properties are of importance for the evaluation of the technical applicability of lignin products. For theoretical considerations, see Sections 3.2.2 and 3.2.3.

4.4.1 Molecular Weight and Polydispersity

Because of difficulties to isolate native lignin from wood without degradation, the data obtained on its molecular weight and polydispersity are often unreliable. Another problem is aggregation of lignin molecules in most solvents, which prevents determination of the real molecular weight. The weight-average molecular weight for MWL from softwood is about 20,000 (Table 4-6), whereas lower values have been reported for hardwood lignin. Compared with cellulose and its derivatives the polydispersity of lignin is high amounting to about 2.5 for MWL from softwood. It has been observed that the polydispersity of dehydrogenase polymers of coniferyl alcohol (synthetic lignin) is increased with increasing average molecular weight and is of the same magnitude (3.0) as for MWL from softwood at $\overline{M}_w$ 20,000.

4.4.2 The Configuration of Dissolved Lignin Polymers

Compared with cellulose, lignin gives solutions of low viscosity, suggesting a compact, spherical structure for the dissolved lignin molecules. As can be seen from Table 3-2 the intrinsic viscosity of lignin is only about $1/_{40}$ of

TABLE 4-6. Molecular Weight and Polydispersity of MWL
Preparations and Dehydrogenase Polymers (DHP) of
Coniferyl Alcohol[a]

Preparation	$\bar{M}_w$	$\bar{M}_w/\bar{M}_n$
MWL		
Eastern spruce	20,600	2.6
Western hemlock	22,700	2.4
DHP	3,700	1.5
	11,000	2.2
	31,400	3.7

[a] From Obiaga (1972).

that of polysaccharides and ¼ of synthetic linear polymers. The exponent *a*
according to the Mark–Houwink equation (p. 58) ranges for lignin from 0.1 to
0.5, corresponding to an intermediate form between an Einstein sphere and
a compact coil (Table 3-4).

References

Adler, E. (1977). Lignin chemistry—Past, present and future. *Wood Sci. Technol.* **11,** 169–218.
Eriksson, Ö., and Lindgren, B. O. (1977). About the linkage between lignin and hemicelluloses in wood. *Sven. Papperstidn.* **80,** 59–63.
Fergus, B. J., and Goring, D. A. I. (1970). The distribution of lignin in birch wood as determined by ultraviolet microscopy. *Holzforschung* **24,** 118–124.
Fergus, B. J., Procter, A. R., Scott, J. A. N., and Goring, D. A. I. (1969). The distribution of lignin in sprucewood as determined by ultraviolet microscopy. *Wood Sci. Technol.* **3,** 117–138.
Freudenberg, K., and Neish, A. C. (1968). "Constitution and Biosynthesis of Lignin." Springer-Verlag, Berlin and New York.
Goring, D. A. I. (1962). The physical chemistry of lignin. *Pure Appl. Chem.* **5,** 233–254.
Goring, D. A. I. (1971). Polymer properties of lignin and lignin derivatives. *In* "Lignins" (K. V. Sarkanen and C. H. Ludwig, eds.), pp. 695–768. Wiley (Interscience), New York.
Higuchi, T., Shimada, M., Nakatsubo, F., and Tanahashi, M. (1977). Differences in biosynthesis of guaiacyl and syringyl lignins in woods. *Wood Sci. Technol.* **11,** 153–167.
Lai, Y. Z., and Sarkanen, K. V. (1971). Isolation and structural studies. *In* "Lignins" (K. V. Sarkanen and C. H. Ludwig, eds.), pp. 165–240. Wiley (Interscience), New York.
Obiaga, T. I. (1972). Lignin molecular weight and molecular weight distribution during alkaline pulping of wood. Ph.D. Thesis, Univ. of Toronto, Toronto.
Sarkanen, K. V. (1971). Precursors and their polymerization. *In* "Lignins" (K. V. Sarkanen and C. H. Ludwig, eds.), pp. 95–163. Wiley (Interscience), New York.
Sarkanen, K. V., and Hergert, H. L. (1971). Classification and distribution. *In* "Lignins" (K. V. Sarkanen and C. H. Ludwig, eds.), pp. 43–94. Wiley (Interscience), New York.
Sarkanen, K. V., and Ludwig, C. H. (1971). Definition and nomenclature. *In* "Lignins" (K. V. Sarkanen and C. H. Ludwig, eds.), pp. 1–18. Wiley (Interscience), New York.
Sarkanen, K. V., and Ludwig, C. H., eds. (1971). "Lignins." Wiley (Interscience), New York.
Wardrop, A. B. (1971). Occurrence and formation in plants. *In* "Lignins" (K. V. Sarkanen and C. H. Ludwig, eds.), pp. 19–41. Wiley (Interscience), New York.

Chapter 5

EXTRACTIVES

Some of the wood constituents, although usually representing only a minor proportion, can be extracted with organic solvents such as ethanol, acetone, or dichloromethane. All major components of softwood resin such as resin acids, fats, and terpenes are removed in this manner. The extracted material contains besides resin a variety of phenolic compounds, for example, flavonoids, lignans, and stilbenes. Certain carbohydrates, tannins, and inorganic salts can be extracted from wood with water, although large amounts of such water-soluble extractives are present only in exceptional cases. However, some trees contain up to 30% tannins and 20–30% arabinogalactan is present in larches.

Sometimes terms *pathological* and *physiological* resin are used. Pathological resin, located in resin canals, is mainly composed of resin acids and monoterpenes and protects the wood against biological damage. Physiological resin, located in the ray parenchyma cells, is rich in fats and constitutes a supply of reserve food. Hardwoods contain only this type of resin.

Extractives are formed through a variety of biosynthetic pathways illustrated for terpenes in Fig. 5-1.

5.1 Softwood Extractives

5.1.1 Resin Canals

Many softwoods contain resin canals. Both vertical and horizontal (radial) canals can exist in the same wood (Fig. 5-2). The resin generated by the epithelial cells surrounding the canal is called oleoresin. In sapwood oleore-

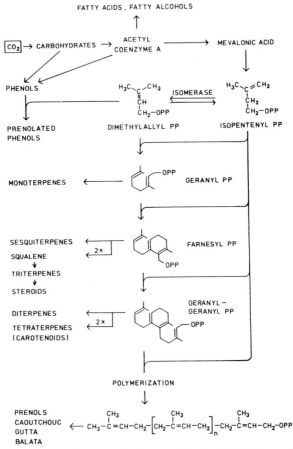

Fig. 5-1. Biosynthesis of terpenes (Lindgren and Norin, 1969). PP, pyrophosphate.

sin in the canals is often under high pressure and can be rapidly exuded at points of injury. The diameter of the resin canals in *Abies, Larix,* and *Picea* is 30 to 100 μm, whereas wider canals are found in *Pinus*—40 to 160 μm, reaching occasionally 300 μm.

About 50% of spruce oleoresin consists of resin acids, 20–30% are volatile monoterpenes, and the remainder, terpenoids and fatty acid esters. Pine oleoresin contains a higher percentage of resin acids, about 70–80%, than spruce oleoresin.

5.1.2 Resin in Parenchyma Cells

More than 95% of the parenchyma cells in softwoods are associated with the wood rays (ray parenchyma). In sapwood, they continue their living

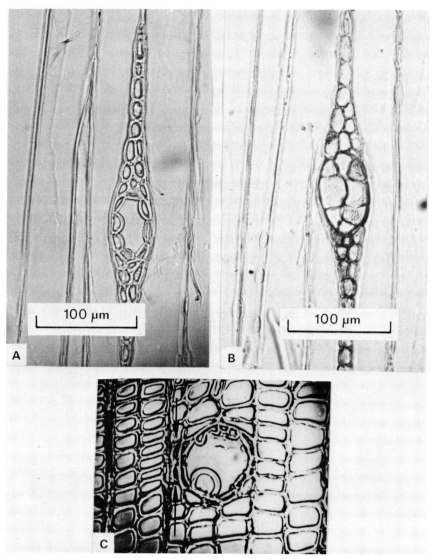

Fig. 5-2. Resin canals in Norway spruce (*Picea abies*) (Back, 1969). (A) Horizontal resin canal in a ray (tangential section) originating from the inner annual rings. The canal is surrounded by epithelial cells which secrete resin into the canal cavities. (B) Horizontal resin canal in a ray (tangential section) originating from the outer annual rings. The canal is filled with epithelial cells because of their swelling during sample preparation. (C) Vertical resin canal (cross section).

functions until the conversion to heartwood takes place. The respiratory activity of living parenchyma cells entails consumption of oxygen and release of carbon dioxide. Tables 5-1 and 5-2 give an idea about the proportions and dimensions of the parenchyma cells in spruce and pine.

The resin in the parenchyma cells is mainly composed of fatty acid esters (fats and waxes) and sterols. When wood is pulped, this resin usually remains encapsulated inside the parenchyma cells, while the oleoresin becomes dispersed in the liquor. This is particularly the case with spruce parenchyma cells, which have minute pores and rigid cell walls. Pine parenchyma cells have larger pores and release their resin more readily (Table 5-1). The resin content of acid sulfite spruce pulps can be effectively lowered by fiber fractionation. The situation is different for pine pulps in which the proportion of parenchyma cells is lower.

5.1.3 Heartwood Extractives

With the death of the living parenchyma cells formation of heartwood begins, and many chemical changes take place. As a consequence, large

TABLE 5-1. Data on Resin, Parenchyma Cells, and Ray Tracheids of Scots Pine (*Pinus sylvestris*) and Norway Spruce (*Picea abies*)[a]

Data	Pine	Spruce
Nonvolatile resin (% of dry wood)	2.2	0.8
Distributed on		
parenchyma cells (%)	30	55
resin canals (%)	70	45
Ray cells in unbleached sulfite and sulfate pulps (weight %)	8	6
Distributed on		
parenchyma cells (%)	30[b]	>85
ray tracheids (%)	70	<15
Mean frequency of parenchymatous to tracheidal ray cells	1:2.8	2.7:1
Mean relative pit area of ray parenchyma cells (% of total cell area)	50[c]	5
Mean pit size at radial walls of parenchyma cells (μm)	12 × 31[d]	2-3 (diam.)

[a] From Back (1969).
[b] One fifth of which contains thin-walled cells.
[c] Total mean value comprising the area of thin-walled cells.
[d] For cells with secondary walls.

TABLE 5-2. Characteristics of Parenchyma Cells in Softwoods[a]

Softwood species	Ray cells		Mean pit size of parenchyma cells (μm)
	Volume % of wood	Parenchyma cells (%)	
Picea	4–6	75–80	2–3 (diam.)
Larix	9–11	80	3 (diam.)
Abies	6–10	85–90	
Thuja	5	97	
Pinus,			
Diploxylon	5–6	70	(10×36)[b]
Haploxylon	6–12	50–60	

[a] From Back (1969).
[b] P. sylvestris.

amounts of extractives are generated, which penetrate throughout the heartwood, including the tracheids. Synthesis of specific fungicides and phenolic substances is typical of pine wood. The extractive content rises from about 4% to 12–14% in pine species (cf. Fig. 5-9).

5.2   Hardwood Extractives

The hardwood resin is located in the ray parenchyma cells which are connected with the vessels. It consists of fats, waxes, and sterols. The accessibility of the resin depends on the pore dimensions as well as on the mechanical stability of the ray parenchyma cells. Considerable variations occur among different wood species (Table 5-3). For instance, the accessibility of the resin in birch is much lower than in aspen.

5.3   The Chemistry of Extractives

The content of extractives and their composition vary greatly among different wood species and also within the different parts of the same tree (cf. Appendix). Wood extractives can be divided into three subgroups: aliphatic compounds (mainly fats and waxes), terpenes and terpenoids, and phenolic compounds. Parenchyma resin is rich in aliphatic components and the oleoresin is mainly composed of terpenoids. Characteristic of the heartwood is the accumulation of phenolic compounds.

TABLE 5-3. Characteristics of Parenchyma Cells in Hardwoods and the Formation of Heartwood[a]

| Genus or species | Volume % of wood | | Average pit size (μm) | Amount in pulps (weight %) | Reason for heartwood formation |
	Ray parenchyma cells	Vertical parenchyma cells			
Betula verrucosa[b] (birch)	9–12	2	1–2	3–10	Secretion
Acer (maple)	12–18		6–10		Secretion
Populus tremula (aspen)	10–11	Very little	8–10	2	Tylose
Fagus sylvatica (beech)	20–22	5	17–20	14	Tylose
Quercus species (oak)	25–40	5	17–20	1	Tylose
Eucalyptus species			30	5	Tylose

[a] From Back (1969).
[b] Cell length 25–50 μm, breadth 20–30 μm.

5.3.1   Aliphatic Components (Fats and Waxes)

A large variety of aliphatic compounds exist in the resin as shown in Table 5-4. The amounts of alkanes and alcohols are relatively small, arachinol (C_{20}), behenol (C_{22}), and lignocerol (C_{24}) representing the major alcohol components. Compounds of this type are very lipophilic and stable.

The fatty acids occur mostly as esters and are the major components of the parenchyma resin in both softwoods and hardwoods. The most important esters are *fats* (glycerol esters), usually present as triglycerides. Esters of other alcohols, which usually are aliphatic alcohols or of terpenoid nature, are known as *waxes*.

The fatty acids are either saturated or unsaturated (Table 5-5). The latter acids, especially the polyunsaturated and conjugated types are quite unstable and readily participate in addition reactions or are easily oxidized. Linoleic acid is a common representative together with oleic and linolenic acids.

5.3.2   Terpenes and Terpenoids

The oleoresin present in the resin canals of certain conifers, especially pine, is secreted as a viscous fluid when the tree is wounded. Pine oleoresin contains about 25% volatile components known as "volatile oil" (or turpentine); the nonvolatile residue consists mainly of resin acids.

TABLE 5-4.   Examples of Aliphatic Extractives in Xylem and Bark[a]

Group	Structure	
n-Alkanes	$CH_3—(CH_2)_n—CH_3$	$n = 8\text{-}30$
Fatty alcohols	$CH_3—(CH_2)_n—CH_2OH$	$n = 16\text{-}22$
Fatty acids	$CH_3—(CH_2)_n—COOH$	$n = 10\text{-}24$
Fats	$CH_2—OR$	R, R', and R"
(glycerol esters)	$\mid$	can be fatty
	$CH—OR'$	acid residues
	$\mid$	(alkyl—CO—)
	$CH_2—OR''$	or hydrogen
		(mono-, di-, and
		triglycerides)
Waxes	$RO—(CH_2)_n—CH_3$	R is fatty acid
(esters of other alcohols)	RO—sterol	residue
	RO—terpene alcohol	
Suberin	Characteristic:	$n = 18\text{-}28$
(polyestolides)	$[—O—(CH_2)_n—CO—]$	
	$[—O—(CH_2)_n—O—CO—(CH_2)_n—CO—]$	

[a] From Lindgren and Norin (1969).

TABLE 5-5. The Most Abundant Fatty Acids of Wood[a]

Acid	Structure
Saturated	
Lauric	$C_{11}H_{23}COOH$
Myristic	$C_{13}H_{27}COOH$
Palmitic	$C_{15}H_{31}COOH$
Stearic	$C_{17}H_{35}COOH$
Arachidic	$C_{19}H_{39}COOH$
Behenic	$C_{21}H_{43}COOH$
Lignoceric	$C_{23}H_{47}COOH$
Unsaturated	
Palmitoleic	$C_{15}H_{29}COOH$ Δ^9
Oleic	$C_{17}H_{33}COOH$ Δ^9
Linoleic	$C_{17}H_{31}COOH$ $\Delta^{9,12}$
Linolenic	$C_{17}H_{29}COOH$ $\Delta^{9,12,15}$
Eleostearic	$C_{17}H_{29}COOH$ $\Delta^{9,11,13}$
Pinolenic	$C_{17}H_{29}COOH$ $\Delta^{5,9,12}$

[a] From Lindgren and Norin (1969).

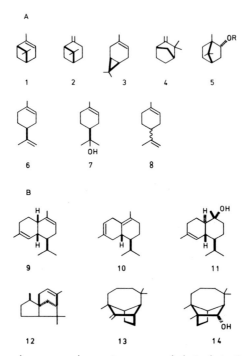

Fig. 5-3. Examples of mono- and sesquiterpenes and their derivatives in softwood. 1, α-Pinene; 2, β-pinene; 3, 3-carene; 4, camphene; 5, borneol (R = H), bornyl acetate (R = $COCH_3$); 6, limonene; 7, α-terpineol; 8, dipentene; 9, α-muurolene; 10, δ-cadinene; 11, α-cadinol; 12, α-cedrene; 13, longifolene; 14, juniperol.

Both the constituents of the volatile oil and the resin acids are of terpenoid nature and are accordingly named terpenoids. Terpenes can formally be considered as condensation products of two or several isoprene (2-methylbutadiene) molecules resulting in dimers and higher oligomers with the elementary formula $(C_{10}H_{16})_n$. The terpenes are divided into monoterpenes, $C_{10}H_{16}$ ($n = 1$), sesquiterpenes $C_{15}H_{24}$ ($n = 1.5$), diterpenes $C_{20}H_{32}$ ($n = 2$), triterpenes $C_{30}H_{48}$ ($n = 3$), tetraterpenes $C_{40}H_{64}$ ($n = 4$), and polyterpenes ($n > 4$). The terpenoids including the polyprenols contain characteristic groups of various types, such as hydroxyl, carbonyl, carboxyl, and ester functions.

The volatile oil of conifers and the turpentine recovered in the kraft pulp industry (see Section 10.3.1) contain monoterpenes and their hydroxy de-rivatives. Minor amounts of sesquiterpenes are also present (Fig. 5-3). Com-pounds of these types are also abundant in needles, bark, and roots.

Diterpenes and their derivatives, which are present in softwood resin, can be grouped into acyclic, monocyclic, dicyclic (Fig. 5-4), and tricyclic (Fig. 5-5) structural types. Since many of these are polyunsaturated, they may readily polymerize to form sparingly soluble high molecular products which give rise to pitch problems in pulping and paper making. The resin acids present in the oleoresin of coniferous woods are derivatives of tricyclic diterpenes. They may be classified into two types: pimaric type, charac-terized by both methyl and vinyl substituents at the C-7 position; and the abietic type, bearing only a single isopropyl group at this position. Because

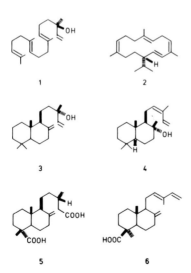

Fig. 5-4. Examples of diterpenes and their derivatives in softwood. 1, Geranyl linalool; 2, thunbergene; 3, β-epimanol; 4, abienol; 5, pinifolic acid; 6, elliotinoic acid (communic acid).

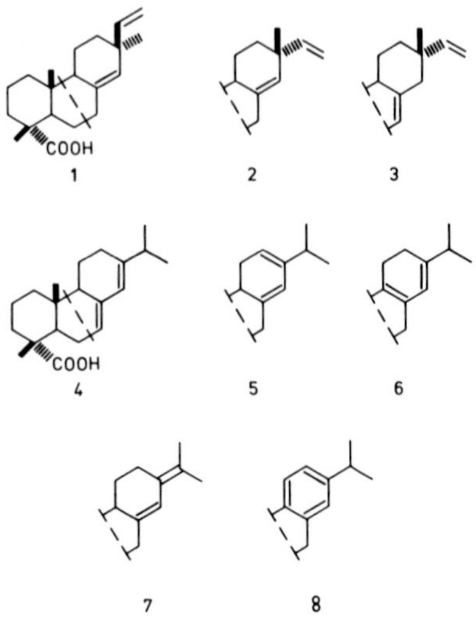

Fig. 5-5.   Examples of resin acids. 1, Pimaric acid; 2, sandaracopimaric acid; 3, isopimaric acid; 4, abietic acid; 5, levopimaric acid; 6, palustric acid; 7, neoabietic acid; 8, dehydroabietic acid.

of their conjugated double bonds, the resin acids of the abietic type are more reactive in isomerization, oxidation, and addition reactions than the pimaric type analogues. Resin and fatty acids form the principal constituents of tall oil, an important by-product in the kraft pulp industry (see Section 10.3.1).

Because of the hydrophobic skeleton in combination with a hydrophilic carboxyl group, the resin acids are good solubilizing agents and contribute effectively (together with the fatty acid soaps) to the removal of resinous substances during kraft pulping and subsequent washing.

Triterpenoids occur in hardwood parenchyma resin, and closely related sterols are also present in softwoods (Fig. 5-6). Sterols typefied by the abundant β-sitosterol, mostly have a hydroxyl group in the C-3 position. They also appear as the alcohol component in fatty acid esters (waxes). Triterpenoids and sterols are sparingly soluble substances contributing to pitch problems in pulping and paper making. Some trees contain polyterpenes and their derivatives known as polyprenols. Betulaprenols, present in birch wood, belong to this category of substances (Fig. 5-7).

5.3.3   Phenolic Extractives and Related Constituents

Structures of the most common phenolic extractives and related constituents are depicted in Fig. 5-8. They constitute a heterogeneous class of compounds, which may be divided into the following groups: (1) *Hydrolyz-*

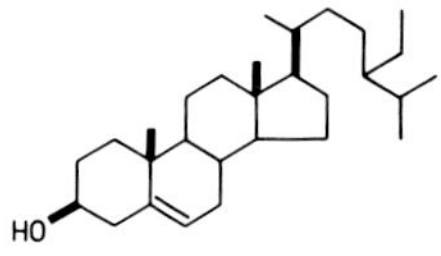

Fig. 5-6. Examples of sterols and triterpenoids in xylem and bark. 1, β-Sitosterol; 2, betulinol; 3, serratenediol; 4, cycloartenol; 5 tremulone.

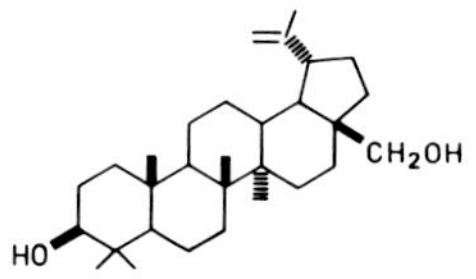

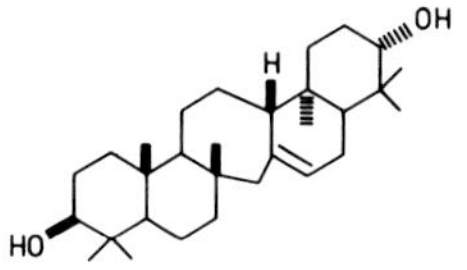

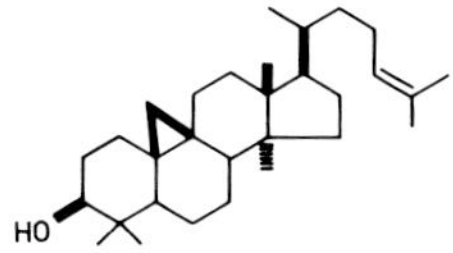

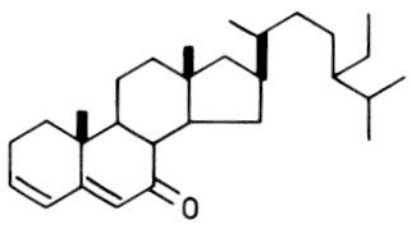

able tannins are a group of substances, which upon hydrolysis yield gallic and ellagic acids and sugars, usually glucose, as main products. Tannins of this type are not very common in woods. (2) Flavonoids are polyphenols, which have a $C_6C_3C_6$ carbon skeleton. Their polymers are called *condensed tannins*. Typical representatives of monomeric flavonoids are chrysin (5,7-dihydroxyflavone) present in Haploxylon pines, and taxifolin (dihy-

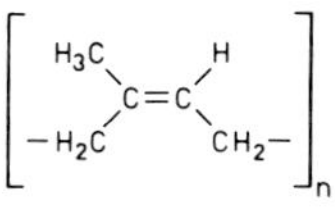

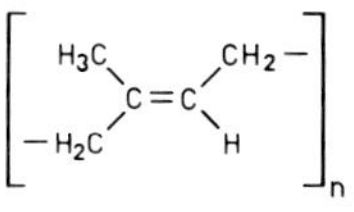

Fig. 5-7. Some structural features of polyprenols in xylem and bark. 1, Caoutchouc (*cis*); 2, balata (*trans*); 3, betulaprenols (60% of the double bonds have *cis*-form).

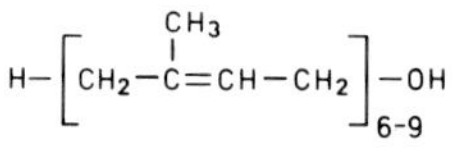

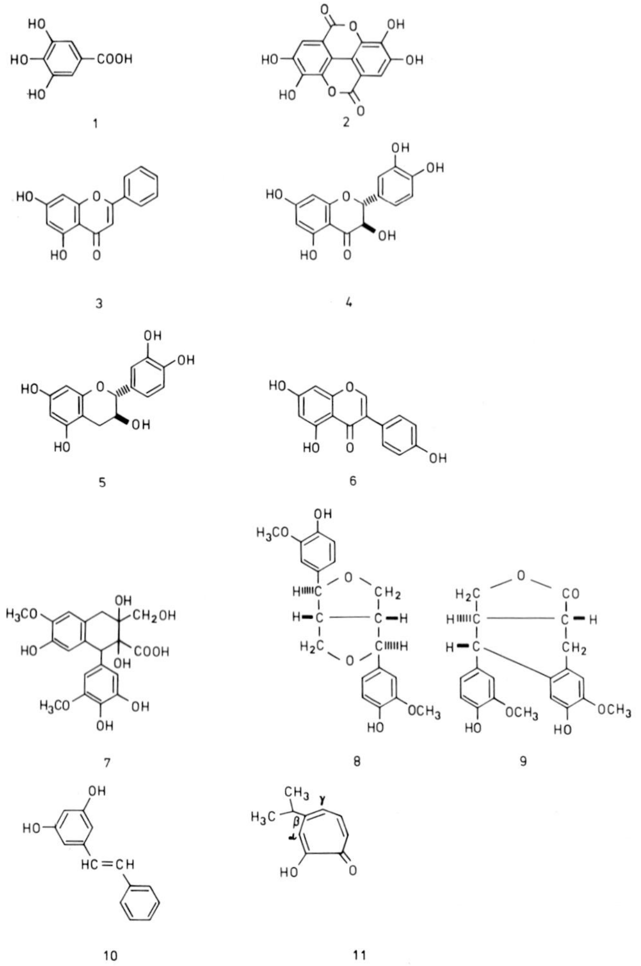

Fig. 5-8. Examples of phenolic extractives and related constituents. 1, Gallic acid; 2, ellagic acid; 3, chrysin; 4, taxifolin; 5, catechin; 6, genistein; 7, plicatic acid; 8, pinoresinol; 9, conidendrin; 10, pinosylvin; 11, β-thujaplicin.

droquercetin), which was originally isolated from Douglas fir heartwood and is present in the cork fraction of inner bark. It is also a common constituent in *Larix* species. The major source of the condensed tannins of the catechin type are quebracho and chestnut wood and wattle bark, but these polyphenols occur also in many other barks belonging to species such as *Eucalyptus* and *Betula*. (3) *Lignans* are formed by oxidative coupling of two phenylpropane (C_6C_3) units, e.g., conidendrin, matairesinol, pinoresinol, and syringaresinol. Lignans related to conidendrin are present in hemlock and spruce species, whereas western red cedar (*Thuja plicata*) contains lignans derived from plicatic acid. (4) *Stilbene* (1,2-diphenylethylene) derivatives possess a conjugated double bond system and are therefore extremely reactive compounds. Pinosylvin, present in *Pinus* species, is an important representative of this group. (5) *Tropolones*, characterized by an unsaturated seven-membered carbon ring, are typical in many decay-resistant conifers, such as cedars, belonging to the family Cupressaceae. For example, α, β-, and γ-thujaplicin have been isolated from western red cedar (*Thuja plicata*) heartwood.

Although the phenolic substances are concentrated in heartwood and bark and only traces are present in xylem, they have fungicidal properties and thus effectively protect the tree against microbiological attack. They also contribute to the natural coloring of wood. However, many of these compounds, especially pinosylvin and taxifolin, are very harmful because they effectively inhibit the delignification by acid sulfite pulping even when present in low concentrations (see Sections 7.2.5, 7.2.7, and 7.2.8). Tropolones form strong complexes with heavy metal ions, such as ferric ions, and can cause corrosion problems in pulping.

The biosynthesis of extractives is controlled genetically and hence each wood species tends to produce specific substances. As a result of secondary changes, heartwood contains a large variety of phenolic substances. From the chemotaxonomical point of view, chemical structures of various flavonoids, lignans, stilbenes, and tropolones are of interest. For example, species within genera such as *Pinus, Acacia,* and *Eucalyptus* can be classified according to their characteristic composition of phenolic substances.

5.4 Variations in Resin Content and Composition

5.4.1 Variations within the Stem

The resin content of wood and its composition vary considerably, depending on such factors as place of growth, age of the tree, and genetic factors. For example, the resin content of Norway spruce (*Picea abies*) is considerably higher for stems grown in the northern than in the southern parts of Scandinavia. The resin content within the same stem also varies, but in a very irregular manner. In all pines, the heartwood contains much more resin than the sapwood, whereas the opposite seems to be true for *Picea* species as indicated by data for Norway spruce. The heartwood extractives in both pine and spruce contain resin acids and free fatty acids as main components,

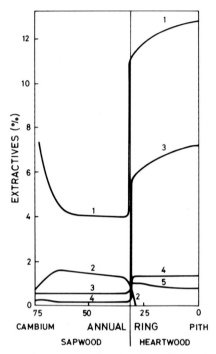

Fig. 5-9. Radial variations in the content and composition of extractives in Scots pine (*Pinus sylvestris*) (Assarsson, 1969; see also Lindgren and Norin, 1969). 1, Total extractives; 2, triglycerides; 3, resin acids; 4, fatty acids; 5, pinosylvin + monomethyl ether.

thus resembling oleoresin in composition. Figure 5-9 illustrates how the resin content and composition vary across the stem of a pine tree.

5.4.2 Changes Caused by Wood Storage

After felling of a tree the content of extractives decreases, and their composition changes because of various reactions. For acid sulfite pulping, the changes are beneficial, and wood is usually stored for several months to minimize pitch problems and to lower the resin content of the pulp. In the case of kraft pulping, storage of wood has detrimental consequences, since yields of both turpentine and tall oil are reduced.

The resin reactions involve both oxidation by air and enzymic hydrolysis. These processes proceed simultaneously, and fats and waxes are mainly hydrolyzed enzymically. Because of both hydrolysis and oxidation the hydrophilicity of the resin constituents is increased. These reactions are largely influenced by the conditions during storage. For example, logs are better preserved when submerged in water than when stored on land. The resin

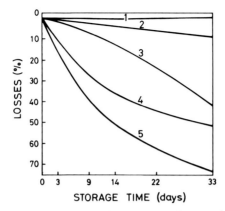

Fig. 5-10. Changes in the resin composition during storage of spruce chips (Assarsson, 1969). 1, Resin acids; 2, unsaponificable substances; 3, extractive content; 4, neutral resin components; 5, esterified acids.

reactions are accelerated if wood is stored in the form of chips. Initially, lingering life processes of ray cells are involved (respiration), changing gradually to degradation by microorganisms, as they invade the wood.

As a consequence of microbiological degradation, wood polysaccharides are attacked during long storage, resulting in reduced pulp yields. This detrimental process can be at least partly avoided by spraying chips with fungicides which, however, also retard the resin reactions. Figure 5-10 illustrates the changes in the resin composition during storage of spruce chips.

References

Assarsson, A. (1969). Changes in resin during wood storage. *Sven. Papperstidn.* **72,** 304–311. (In Swed.)

Back, E. (1969). Wood-anatomical aspects on resin problems. *Sven. Papperstidn.* **72,** 109–121. (In Swed.)

Browning, B. L. (1967). "Methods of Wood Chemistry," Vol. 1 Wiley (Interscience), New York.

Clayton, R. B. (1970). The chemistry of nonhormonal interactions: Terpenoid compounds in ecology. *In* "Chemical Ecology" (E. Sondheimer and J. B. Simeone, eds.), pp. 235–280. Academic Press, New York.

Erdtman, H. (1973). Molecular taxonomy. *In* "Phytochemistry" (L. P. Miller, ed.), Vol. 3, pp. 327–350. Van Nostrand-Reinhold, New York.

Hillis, W. E., ed. (1962). "Wood Extractives and Their Significance to the Pulp and Paper Industry." Academic Press, New York.

Kimland, B., and Norin, T. (1972). Wood extractives of common spruce, *Picea abies* (L.) Karst. *Sven. Papperstidn.* **75,** 403–409.

Lindgren, B., and Norin, T. (1969). The chemistry of resin. *Sven. Papperstidn.* **72,** 143–153. (In Swed.)

Miller, L. P., ed. (1973). "Phytochemistry," Vol. 2. Van Nostrand-Reinhold, New York.

Chapter 6

BARK

Bark is the layer external to the cambium which surrounds the stem, branches, and roots, amounting to about 10-15% of the total weight of the tree. Debarked wood is normally used for pulping and even traces of bark residues detrimentally affect the pulp quality. The resulting bark waste is usually burned under recovery of heat. Despite extensive studies only a small fraction of bark is used today as raw material for production of chemicals (see Section 10.1).

6.1 Anatomy of Bark

Bark is composed of several cell types and its structure is complicated in comparison with wood. In addition to variations occurring within the same species, depending on such factors as age and growth conditions of the tree, each species is characterized by specific features of its bark structure.

Bark can roughly be divided into living inner bark or *phloem* and dead outer bark or *rhytidome*. The tissues of the bark substance are formed either by primary or secondary growth. The primary growth means direct production of embryonal cells at the growing points of the stem apex and their further development to primary tissues. *Epidermis, cortex,* and *primary phloem* are primary tissues (Fig. 6-1). The formation of secondary tissues

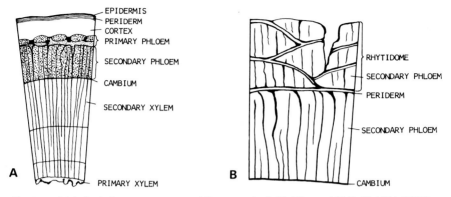

Fig. 6-1. Main bark tissues: young stem (A), mature bark (B) (Chang, 1954). © 1954 *TAPPI.* Reprinted from *Anatomy of Common North American Pulpwood Barks,* TAPPI Monograph 14, with permission.

takes place in two special meristems; in vascular cambium, which produces the secondary phloem; and in the cork cambium (*phellogen*), which generates *periderm.* Continuous division of cells gives rise to several periderm layers. In mature bark the last-formed periderm is the boundary between the inner and outer bark.

6.1.1 Inner Bark

The main components of inner bark are sieve elements, parenchyma cells, and sclerenchymatous cells. *Sieve elements* perform the function for transportation of liquids and nutrients. More specifically and according to their shape the sieve elements are divided into sieve cells and sieve tubes. The former types are present in gymnosperms, the latter in angiosperms. The sieve elements are arranged in longitudinal cell rows which are connected through sieve areas. The sieve cells are comparatively narrow with tapering ends, whereas the sieve tubes are thicker and cylindrical. After 1–2 years, or after a longer time in the monocotyledons, the activity of the sieve elements ceases and they are replaced by new elements.

Parenchyma cells have the function of storing nutrients and are located between the sieve elements in the inner bark. Both vertical parenchyma cells and horizontal phloem rays are present. The latter are direct continuations of the xylem rays, but much shorter.

Sclerenchymatous cells function as the supporting tissue observable in most tree species as layers corresponding to the annual rings in xylem. According to their shape two types are distinguishable: the bast fibers, usually measuring 0.1–3 mm in length and often arranged in tangential rows, and the sclereids or stone cells, which are short and rounded and located as layers between the sieve elements.

6.1.2   Outer Bark

The outer bark, which consists mainly of periderm or cork layers, protects the wood tissues against mechanical damage and preserves it from temperature and humidity variations. In most woody plants a periderm replaces the epidermis within the first year of growth. The first periderm in stems usually arises from the cork cambium in the outer surface of bark, either in the subepidermal layer or in the epidermis. The following periderms are then formed in successively deeper layers of the bark or in the bast tissue. Cork tissue is predominantly formed in the outward direction, but some division also occurs inward resulting in so-called phelloderm tissue resembling parenchyma cells. Owing to this sequence the final rhytidome usually occurs as scaly bark and, in addition to the cork cells, contains the same cells as those present in the bast.

The cork cells, which consist of three thin layers and are only rarely pitted, are arranged in radial rows and die at an early stage. They are cemented together to a tight tissue resisting water and gases. Because of different growth activity in the spring and in the late summer separate layers are formed in the bark corresponding to the annual rings in the xylem.

As a dead tissue the rhytidome cannot expand and accommodate the radial growth of the stem and is therefore crushed. The resulting form of the cracked bark depends on the anatomical structure and elasticity of the rhytidome and is typical of each tree species.

6.2   Chemistry of Bark

The chemical composition of bark is complicated, varies among the different tree species and also depends on the morphological elements involved. Many of the constituents present in wood also occur in bark, although their proportions are different. Typical of bark is its high content of certain soluble constituents (extractives) such as pectin and phenolic compounds as well as suberins. The mineral content of bark is also much higher than that in wood.

Bark can roughly be divided into the following fractions: fibers, cork cells, and fine substance including the parenchyma cells. The fiber fraction is chemically similar to that of the wood fibers and consists of cellulose, hemicelluloses, and lignin. The other two fractions contain large amounts of extractives. The walls of the cork cells are impregnated with suberin, whereas the polyphenols are concentrated in the fine fraction.

6.2.1 Soluble Constituents (Extractives)

Bark extractives can roughly be divided into lipophilic and hydrophilic constituents, although these groups do not have any distinctive boundaries. The total content of both lipophilic and hydrophilic extractives is usually high in bark compared with wood and varies within wide limits among different species, corresponding to 20–40% of the dry weight of bark. These extractives constitute an extremely heterogeneous group of substances some of which are typical of bark but are rarely present in xylem (cf. Chapter 5).

The lipophilic fraction, extractable with nonpolar solvents (ethyl ether, dichloromethane, etc.) consists mainly of fats, waxes, terpenes and terpenoids, and higher aliphatic alcohols (cf. Sections 5.3.1 and 5.3.2). Terpenes, resin acids, and sterols are located in the resin canals present in the bark and also occur in the cork cells and in the pathological exudate (oleoresin) of wounded bark. Triterpenoids are abundant in bark: β-sitosterol occurs in waxes, as an alcohol component, and the cork cells in the outer bark (periderm) of birch contain large amounts of betulinol (cf. Fig. 5–6).

The hydrophilic fraction, extractable with water alone or with polar organic solvents (acetone, ethyl alcohol, etc.) contains large amounts of phenolic constituents (cf. Section 5.3.3 and Fig. 5-8). Many of them, especially the condensed tannins (often called "phenolic acids") can be extracted only as salts with dilute solutions of aqueous alkali. For example, considerable quantities of flavonoids, belonging to the group of condensed tannins, are present in the bark of hemlock, oak, and redwood. Monomeric flavonoids, including quercetin and dihydroquercetin (taxifolin), are also present in bark. Small amounts of lignans and stilbenes (e.g., piceatannol in spruce bark) occur as well. Glycosides of simple plant phenols, such as salicin and coniferin are present in barks of *Salix* and *Picea* species, respectively. Compounds belonging to the extremely heterogeneous group of hydrolyzable tannins are further phenolic constituents occurring in bark. Because the ester linkages in these tannins are partly hydrolyzed even in warm water, the resulting insoluble ellagic and gallic acids are readily precipitated (cf. Fig. 5-8).

Minor amounts of soluble carbohydrates, proteins, vitamins, etc., are present in the bark. In addition to starch and pectins, oligosaccharides, including raffinose and stachyose have been detected in phloem exudates.

6.2.2 Insoluble Constituents

Polysaccharides, lignin, and suberins are the principal cell wall constituents of bark.

Polysaccharides. The bast fibers are essentially built up by polysaccharides. Cellulose dominates (roughly 30% of the dry bark weight) in addition to the hemicelluloses, which are of the same type as in wood (see Section 3.3).

In addition, a highly branched arabinan probably occurs in many barks, and especially pines. The connecting strands of the sieve elements are surrounded by a polysaccharide called callose, which is a $(1 \rightarrow 3)$-β-D-glucan.

Lignin. No completely satisfactory data are available on the lignin in bark because of the difficulties to separate it from the phenolic acids. Lignin contents of about 15–30% (based on extractive-free bark weight) have been reported for coniferous bark derived from different wood species. Other studies indicate that inner bark lignin is similar to wood lignin, whereas the outerbark lignin significantly differs from it. Further work is needed, however, to confirm these differences.

Suberins. The cork cells in the outer bark contain polyestolides or suberins. The suberin content in the outer layer of the cork oak bark (cork) is especially high and amounts to 20–40% in the periderm of birch bark. Polyestolides are complicated polymers composed of ω-hydroxy monobasic acids which are linked together by ester bonds. In addition, they contain α,β-dibasic acids esterified with bifunctional alcohols (diols) as well as ferulic and sinapic acid moieties. The chain lengths vary but suberins are enriched with molecules having 16 and 18 carbon atoms. There are also double bonds and hydroxyl groups through which ester and ether cross-links are possible. The outer layer of the epidermis contains so-called cutin, which is heavily branched and has a structure similar to suberin.

6.2.3 Inorganic Constituents

Bark contains 2–5% inorganic solids of the dry bark weight (determined as ash). The metals are present as various salts including oxalates, phosphates, silicates, etc. Some of them are bound to the carboxylic acid groups of the bark substance. Calcium and potassium are the predominating metals. Most of the calcium occurs as calcium oxalate crystals deposited in the axial parenchyma cells. Bark also contains trace elements, such as boron, copper, and manganese.

References

Browning, B. L. (1967). "Methods of Wood Chemistry," Vol. 1. Wiley (Interscience), New York.
Chang, Y. (1954). "Anatomy of Common North American Pulpwood Barks," TAPPI Monograph Series, No. 14. TAPPI, New York.

Jensen, W., Fremer, K. E., Sierilä, P., and Wartiovaara, V. (1963). The chemistry of bark. *In* "The Chemistry of Wood" (B. L. Browning, ed.), pp. 587–666. Wiley (Interscience), New York.

Martin, J. T. (1973). Cutins and suberins. *In* "Phytochemistry" (L. P. Miller, ed.), Vol. 3, pp. 154–161. Van Nostrand-Reinhold, New York.

Srivastava, L. M. (1964). *Int. Rev. For. Res.* **1,** 203–277.

Chapter 7

PULPING CHEMISTRY

7.1 Historical Background

The first patent dealing with pulping of wood with aqueous solutions of calcium hydrogen sulfite and sulfur dioxide in pressurized systems was granted in 1866. This pioneering invention, made in the United States by B. Tilghman, can be considered to be the origin of the sulfite pulping process. It required almost one decade before the world's first sulfite pulp mill started its production in Sweden in 1874. This was accomplished by C. D. Ekman, who is the principal initiator of the sulfite pulp industry.

Essentially, sulfite pulping is still based on these old inventions although several innovative modifications and technical improvements have been introduced. The more recent achievements during the 1950s and the 1960s concern the introduction of soluble bases, i.e., replacement of calcium by magnesium, sodium, or ammonium, enabling increased flexibility in adjusting the cooking conditions and the production of a variety of different pulp types. Also, methods for the recovery of these bases as well as sulfur dioxide have been developed.

Pressurized alkaline cooking systems at high temperatures were introduced in the 1850s. According to the method proposed by C. Watt and H. Burgess, sodium hydroxide solution was used as cooking liquor and the resulting spent liquor was concentrated by evaporation and burned. The smelt, consisting of sodium carbonate, was reconverted to sodium hydroxide by calcium hydroxide (caustisizing). Since sodium carbonate was used for makeup, the cooking process was named the "soda process."

In 1870, A. K. Eaton in the United States patented the use of sodium sulfate instead of sodium carbonate. Similar ideas were pursued by C. F. Dahl, who about 15 years later presented a technically feasible pulping process in Danzig. These inventions initiated the "*sulfate*" or "*kraft*" *process*, which today is by far the most common chemical pulping process. The decisive breakthrough of this process came, however, first in the 1930s after the pioneering work of G. H. Tomlinson in Canada, who developed a recovery furnace suitable for combustion of sulfate spent liquors. The kraft process has almost completely replaced the old soda process because of its superior delignification selectivity resulting also in a higher pulp quality. Since the 1960s the production of kraft pulps has also increased much more rapidly than that of sulfite pulps due to several factors, such as a simpler recovery of chemicals and higher fiber strength. The introduction of effective bleaching agents, and especially chlorine dioxide, has eliminated the earlier difficulties in bleaching of kraft pulps to a high brightness, and prehydrolysis of wood has made it possible to produce high grade dissolving pulps by the kraft process.

The attention in the pulp industry today is focused on developing methods for minimizing pollution and saving energy. An inherent disadvantage of the kraft process is still the unpleasant odor emitted to the surroundings. The water pollution problems caused by bleach plant effluents are still unsolved. It is not surprising, therefore, that increasing interest is directed toward the development of sulfur-free pulping and chlorine-free bleaching processes. This need, together with the need for the development of novel chemical products from wood and lignocellulosic waste material, and for expanding the use of wood and fibers by new modification processes, will offer a great challenge to wood chemistry in the future.

7.2 Sulfite Pulping

7.2.1 Cooking Chemicals and Equilibria

Sulfur dioxide is a two-basic acid and the following equilibria prevail in its aqueous solution:

$$SO_2 + H_2O \rightleftarrows H_2SO_3 \ (SO_2 \cdot H_2O) \tag{7-1}$$

$$H_2SO_3 \rightleftarrows H^+ + HSO_3^- \tag{7-2}$$

$$HSO_3^- \rightleftarrows H^+ + SO_3^{2-} \tag{7-3}$$

Since the concentrations of sulfur dioxide in its free (SO_2) and hydrated ($SO_2 \cdot H_2O$ or H_2SO_3) forms cannot be determined separately, equations (7-1) and (7-2) are combined to give expression (7-4) in which the total sulfur dioxide concentration is in the dominator. This defines the first equilibrium constant K_1:

$$K_1 = [H^+] [HSO_3^-] / ([SO_2] + [H_2SO_3]) \qquad (7\text{-}4)$$

The second equilibrium constant derived from equation (7-3) is

$$K_2 = [H^+] [SO_3^{2-}] / [HSO_3^-] \qquad (7\text{-}5)$$

By taking the logarithms of both sides of equations 7-4 and 7-5, the following expressions are obtained:

$$pK_1 = pH - \log([HSO_3^-] / ([SO_2] + [H_2SO_3]) \cong 2 \qquad (7\text{-}6)$$

$$pK_2 = pH + \log([HSO_3^-] / [SO_3^{2-}]) \cong 7 \qquad (7\text{-}7)$$

It should be noted that K_1 and K_2 are not thermodynamic constants since the activity coefficients have been neglected, and hence they are strictly valid only at a given concentration.

It follows from these equilibria that the relative concentrations of sulfur dioxide, hydrogen sulfite, and sulfite are governed by the pH of the solution (Fig. 7-1). As can be seen, sulfur dioxide is present almost exclusively in the form of hydrogen sulfite ions at pH around 4. Below and above this value the concentrations of sulfur dioxide and sulfite ions, respectively, are successively increased. These equilibria also vary with the temperature. At temperatures used for pulping (130°–170°C), the actual pH value is higher than that

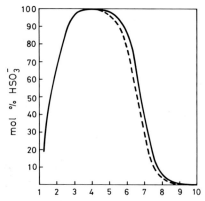

Fig. 7-1. The molar ratio of hydrogen sulfite ion concentration to total sulfur dioxide as a function of pH at 25°C (Sjöström et al., 1962). ——, 10 g Na_2O/liter; - - -, 50 g Na_2O/liter. $pK_1 \sim$ 1.7; $pK_2 \sim$ 6.6-6.8.

measured at room temperature and this deviation is larger in the acidic region.

Because of the low solubility of calcium sulfite, a large excess of sulfur dioxide is required to avoid its formation from calcium hydrogen sulfite. Calcium base is thus usable only for acid sulfite pulping. Magnesium sulfite is much more soluble. When using magnesium as base, the pH can be increased to about 4–5, but above this range magnesium sulfite starts to precipate and in alkaline region magnesium precipates as hydroxide. Sodium and ammonium sulfites and hydroxides are easily soluble, and the use of these bases have no limitations in the pH of the cooking liquor.

There are several modifications of the sulfite method which are designated according to the pH of the cooking liquor (Table 7-1). For the production of chemical pulps, delignification is allowed to proceed until most of the lignin in the middle lamella is removed after which the fibers can be readily separated from each other. Semichemical pulps are often produced by the neutral sodium sulfite method followed by mechanical fiberization of the partially delignified wood.

According to the usual but rather misleading convention, the total amount of sulfur dioxide is divided into "free" and "combined" sulfur dioxide. For example, sodium hydrogen sulfite solution contains equal amounts of combined and free sulfur dioxide ($2 NaHSO_3 \rightarrow Na_2SO_3 + SO_2 + H_2O$), although essentially no free sulfur dioxide exists in such a solution. The term "active base" refers to the sum of the hydrogen sulfite and sulfite ions and is usually expressed as oxide, e.g., CaO or Na_2O. A typical acid sulfite cooking liquor contains about 10 g and 60 g combined and free sulfur dioxide per liter, respectively.

7.2.2 Impregnation

The cooking process begins with an impregnation stage after the chips have been immersed in the cooking liquor. This stage involves both the liquid *penetration* into wood cavities and the *diffusion* of dissolved cooking chemicals. The rate of penetration depends on the pressure gradient and proceeds comparatively rapidly, whereas diffusion is controlled by the concentration of dissolved chemicals and takes place more slowly. Penetration is influenced both by the pore size distribution and capillary forces while diffusion is regulated only by the total cross-sectional area of accessible pores.

A good impregnation is a prerequisite for a satisfactory cook. If the transport of chemicals into the chips is still incomplete after the cooking temperature has been reached, undesirable reactions catalyzed by hydrogen ions will occur. For instance, if the base concentration in an acid sulfite cook is

TABLE 7-1. Sulfite Pulping Methods and Conditions

Method	pH range	"Base" alternatives	Active reagents	Max. temp. (°C)	Time at max. temp. (hr)	Softwood pulp yield (%)
Acid (bi)sulfite	1-2	Ca^{2+}, Mg^{2+}, Na^+, NH_4^+	HSO_3^-, H^+	125-145	3-7	45-55
Bisulfite	3-5	Mg^{2+}, Na^+, NH_4^+	HSO_3^-, H^+	150-170	1-3	50-65
Two-stage sulfite (Stora type)						
Stage 1	6-8	Na^+	HSO_3^-, SO_3^{2-}	135-145	2-6	50-60
Stage 2	1-2	Na^+	HSO_3^-, H^+	125-140	2-4	
Three-stage sulfite (Sivola type)						
Stage 1	6-8	Na^+	HSO_3^-, SO_3^{2-}	120-140	2-3	
Stage 2	1-2	Na^+	HSO_3^-, H^+	135-145	3-5	35-45
Stage 3	6-10	Na^+	HO^-	160-180	2-3	
Neutral sulfite (NSSC)	5-7	Na^+, NH_4^+	HSO_3^-, SO_3^{2-}	160-180	0.25-3	75-90[a]
Alkaline sulfite	9-13	Na^+	SO_3^{2-}, HO^-	160-180	3-5	45-60

[a] Hardwood.

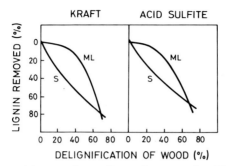

Fig. 7-2. Delignification of the secondary wall (S) and compound middle lamella (ML) during kraft and acid sulfite pulping (Wood and Goring, 1973). Note that the S wall is delignified faster than the ML layer at the earlier stages of the cook.

insufficient the sulfonic acids formed are not neutralized, and the pH value of the cooking liquor drops sharply. Because of the low pH value, reactions leading to lignin condensation as well as to decomposition of the cooking acid are accelerated in the interior of the chips resulting in dark, hard cores. In the choice of the length, pressure and temperature of impregnation, consideration must be given to the wood used; for example, heartwood is much more difficult to impregnate than sapwood.

7.2.3 Morphological Factors Involved in Delignification

An interesting question not yet fully clarified concerns the sequence of delignification of the different parts of the cell wall. Despite of conflicting

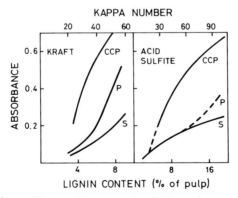

Fig. 7-3. UV absorbance (222 nm, 0.5 μm section thickness) by various morphological regions of spruce fibers delignified to various lignin contents by the kraft and acid sulfite method (Wood and Goring, 1973). S, secondary wall; P, primary wall; CCP, primary wall at the cell corner.

TABLE 7-2.   Distribution of Lignin in Kraft and Acid Sulfite Fibers of Spruce Earlywood[a]

| Pulping method: | Proportion of total lignin in fiber (%) | | | | |
| | Kraft | | Acid sulfite | | |
Kappa number:	50	25	50	25	15
Secondary wall	73	87	88	90	92
Primary wall	14	10	8	8	8
Cell corner	13	3	4	2	0

[a] From Wood and Goring (1973).

opinions it is to be expected that the chemicals diffuse gradually from the lumen through the cell layers reaching the middle lamella at the end. Indeed, this view has been supported by experimental data according to which the delignification during acid sulfite pulping proceeds more rapidly in the secondary wall than in the middle lamella (Fig. 7-2). Toward the end of the cook the differences in lignin concentration in the primary and secondary walls are reduced (Fig. 7-3). It should be noted that the residual lignin remaining in the pulp is located mainly in the secondary wall because of its thickness compared with that of the middle lamella (Table 7-2). The progress of delignification in kraft pulping is analogous.

7.2.4   General Aspects of Delignification

Ideally, the purpose of delignification is to remove the lignin at least from the middle lamella as selectively as possible. In reality, however, attack on the polysaccharides cannot be avoided and this results in depolymerization of polysaccharides and losses of hemicellulose.

The *selectivity* of delignification of acid sulfite pulping is illustrated for softwoods in Fig. 7-4. For comparison, corresponding selectivity curves for two-stage sulfite and kraft processes have also been included. It may be noted that regardless of the pulping method used, a drastic dissolution of carbohydrates occurs at the initial stages of the cook. The selectivity is then gradually improved, but diminishes again toward the end of the cook.

The desirable reactions of lignin during sulfite pulping are of two types: sulfonation and hydrolysis. Sulfonation introduces hydrophilic sulfonic acid groups into the lignin polymer, while hydrolysis breaks ether bonds, creating new phenolic hydroxyl groups and lowering the molecular weight. Both reactions increase the hydrophilicity of the lignin, rendering it more soluble.

Both sulfonation and hydrolysis are necessary for the dissolution of lignin to take place. At higher pH values (pH > 6), hydrolysis is slow and becomes

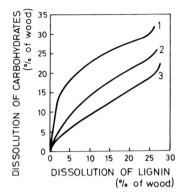

Fig. 7-4. Selectivity of lignin dissolution during (1) kraft, (2) acid sulfite, and (3) two-stage sulfite pulping of softwood (see Sjöström, 1964). Note that the delignification proceeds very unselectively at the beginning and at the end of the cook.

the rate determining step in delignification. At low pH (1–2), hydrolysis becomes fast compared with sulfonation, which assumes the role of the rate-determining step. Diffusion of the cooking chemicals into the reaction zone as well as transport of the reaction products out into the solution also influence the rate of delignification. The base present in the cooking liquor is required for neutralization of the lignosulfonic acids in addition to other acids formed in side reactions. If not enough alkali is present, pH is sharply decreased and the competing condensation reactions prevent delignification (Fig. 7-5).

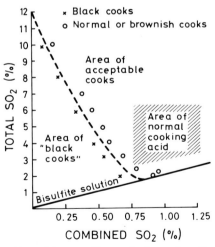

Fig. 7-5. Influence of cooking acid composition on lignin condensation in acid sulfite cooking of spruce (130°C) (Kaufmann, 1951).

7.2.5 Lignin Reactions

At a given temperature, the extent of delignification depends largely on the acidity of the cooking liquor. Conditions typical of acid sulfite pulping (140°C, pH 1–2) result in effective delignification, whereas after a corresponding treatment in neutral sulfite solution most of the lignin remains insoluble. The degree of sulfonation of the undissolved softwood lignin, expressed as the molar ratio of sulfonic acid groups to lignin methoxyl also remains low ($SO_3H/OCH_3 \sim 0.3$). In contrast, the degree of sulfonation of softwood lignosulfonates dissolved during acid sulfite pulping is much higher or about 0.5.

Most of the sulfonic acid groups introduced into the lignin replace hydroxyl or ether functions at the α-carbon atom of the propane side chain. The sulfonation proceeds rapidly at all pH values when the phenolic hydroxyl group located at the *para* position is free. Experiments especially with dimeric model compounds representing various structures and bond types in lignin have provided insight into these reactions. Under acidic conditions the most important structural units in lignin are sulfonated irrespective of whether they are free or etherified. Under neutral conditions, however, sulfonation as well as cleavage of the ether bonds, leading to lignin fragmentation, is essentially restricted to the phenolic units.

Acid Sulfite (and Bisulfite) Pulping During acid sulfite pulping the α-hydroxyl and the α-ether groups are cleaved readily under simultaneous formation of benzylium ions (Fig. 7-6). This reaction takes place regardless

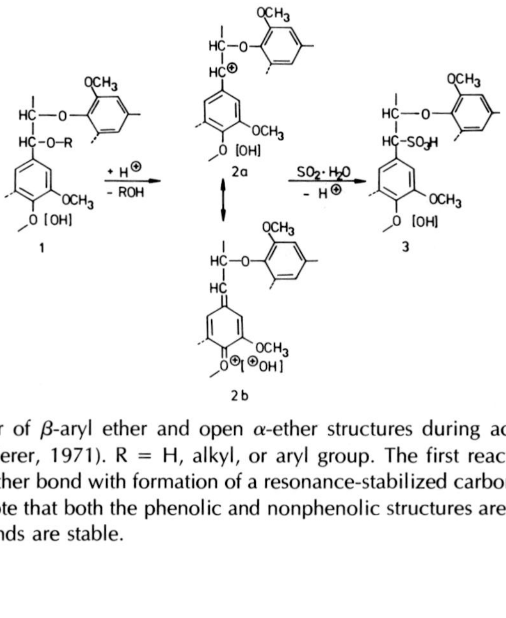

Fig. 7-6. Behavior of β-aryl ether and open α-ether structures during acid sulfite pulping (Gellerstedt and Gierer, 1971). R = H, alkyl, or aryl group. The first reaction step involves cleavage of the α-ether bond with formation of a resonance-stabilized carbonium ion which is then sulfonated. Note that both the phenolic and nonphenolic structures are sulfonated, while the β-aryl ether bonds are stable.

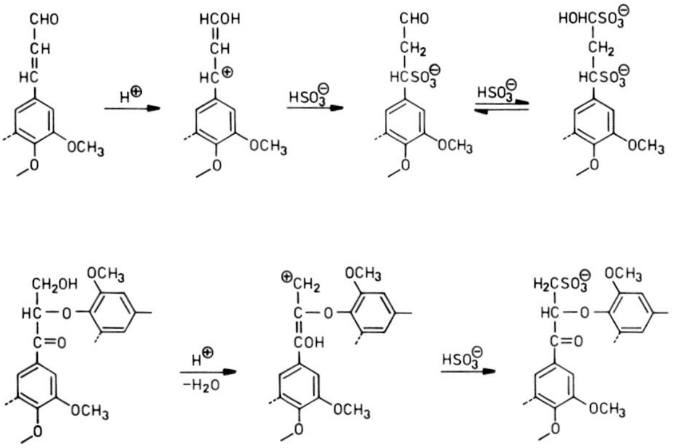

Fig. 7.7. Sulfonation of coniferaldehyde end groups and substituted structures containing α-carbonyl groups (see Gellerstedt, 1976). At lower temperatures aldehyde end groups can bind sulfur dioxide because of the formation of α-hydroxysulfonic acid.

of whether the phenolic hydroxyls of the phenylpropane units are etherified or free. The cleavage of open α-aryl ether bonds represents the only noteworthy fragmentation of lignin during acid sulfite pulping. Although only 6–8% open α-aryl ether bonds are present in softwood lignin, their cleavage results in a considerable fragmentation. The benzylium ions are sulfonated by attack of hydrated sulfur dioxide or bisulfite ions present in the cooking liquor. The coniferaldehyde end groups and β-substituted structures containing α-carbonyl groups are also sulfonated (Fig. 7-7). The benzylium ions formed from the 1,2-diarylpropane structures are easily converted to stilbene structures by elimination of a hydrogen ion at the β-position, after which the electrophilic γ-carbon atoms can be sulfonated (Fig. 7-8).

Condensation reactions of carbonium ions compete with sulfonation and their frequency is increased with increasing acidity. Carbon–carbon bonds are formed most commonly when the benzylium ions react with the weakly nucleophilic 1- and 6-(or 5-)positions of other phenylpropane units (Fig. 7-9). Subsequently, propane side chains and hydrogen ions are eliminated, respectively. In general, the condensation reactions result in increased molecular weight of the lignosulfonates and the solubilization of lignin is

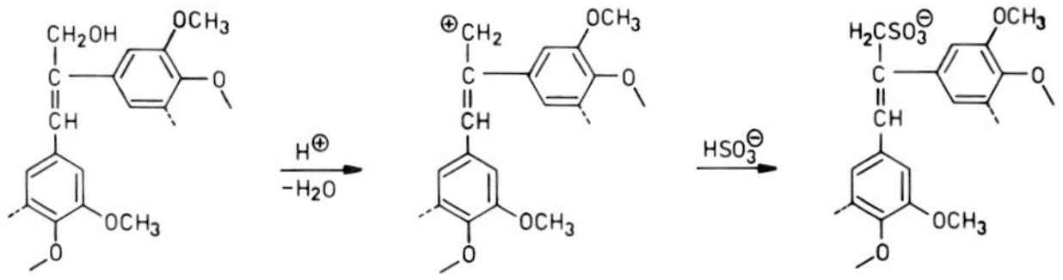

Fig. 7-8. Reactions of stilbene structures during acid sulfite pulping (see Gellerstedt, 1976).

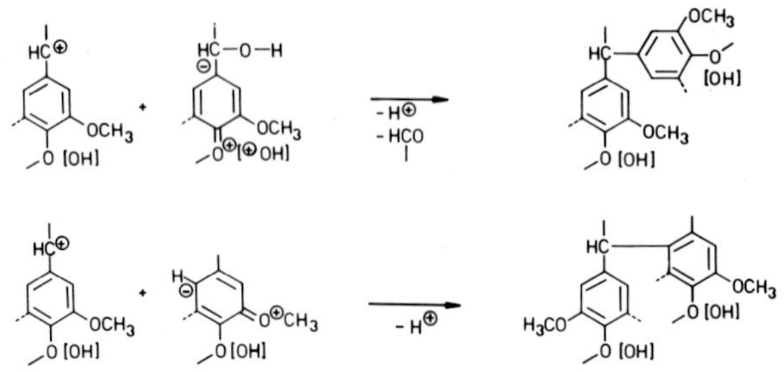

Fig. 7-9.   Examples of lignin condensation products formed during acid sulfite pulping (Gierer, 1970). Condensation results from the reaction of a carbonium ion with the weakly nucleophilic sites in the benzene nucleus.

retarded or inhibited. However, the benzylium ions formed from phenyl coumaran and pinoresinol structures can be condensed intramolecularly without increasing the molecular weight (cf. Fig. 7-10).

During acid sulfite pulping, lignin may also condense with reactive phenolic extractives. Pinosylvin and its monomethyl ether, present in pine heartwood, are examples of phenolic extractives of this type. Dual condensation of pinosylvin with lignin generates harmful cross-links. Consequently, pine heartwood cannot be delignified by the conventional acid sulfite method.

Cross-links between lignin entities may also be generated by thiosulfate present in the cooking liquor (Fig. 7-11). This results in retarded delignification and, under certain circumstances, in complete inhibition ("black cook"). For the formation of thiosulfate, see Section 7.2.8.

Neutral and Alkaline Sulfite Pulping   In neutral sulfite pulping the most important reactions of lignin are restricted to phenolic lignin units only. The first stage always proceeds via the formation of a quinone methide with simultaneous cleavage of an α-hydroxyl or an α-ether group (Fig. 7-12). At least in noncyclic structures, the quinone methide is readily attacked by a

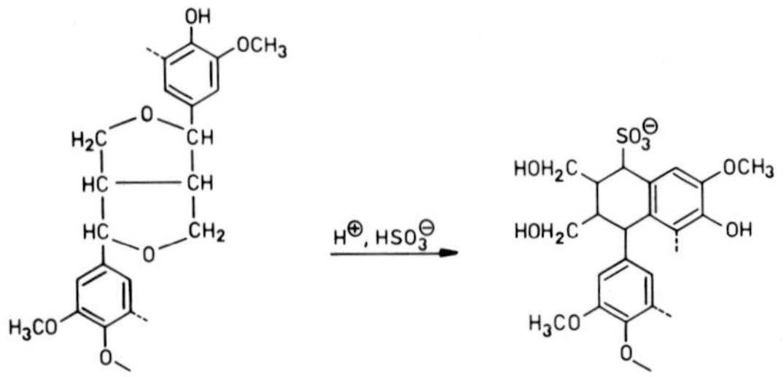

Fig. 7-10.   Intramolecular condensation of pinoresinol structures during acid sulfite pulping (see Gellerstedt, 1976).

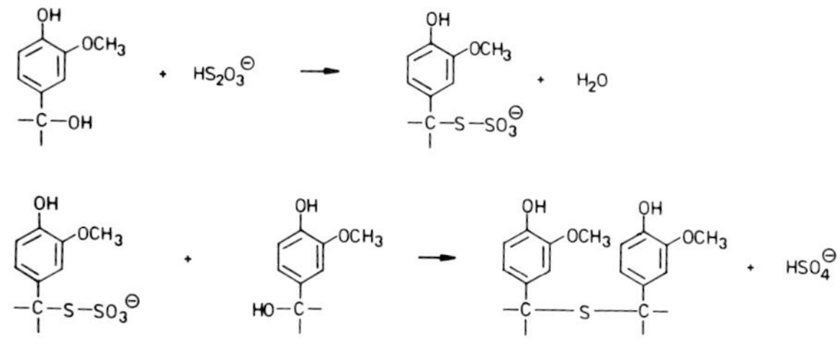

Fig. 7-11.   Reactions of lignin with thiosulfate (see Goliath and Lindgren, 1961).

sulfite or a bisulfite ion. The α-sulfonic acid group formed facilitates the nucleophilic displacement of the β-substituent in β-aryl ether structures by a sulfite or bisulfite ion. Subsequent loss of the α-sulfonate group leads to a styrene-β-sulfonic acid structure, especially at higher pH values (>7). The cleavage of the α- as well as the β-aryl ether bonds naturally generates new reactive phenolic units.

The quinone methides can also react by elimination of formaldehyde or hydrogen ion at the β-carbon atom, especially when the formation of conjugated diaryl structures is possible. Examples of this type of reactions are the formation of stilbenes from phenyl coumarans or 1,2-diarylpropane struc-

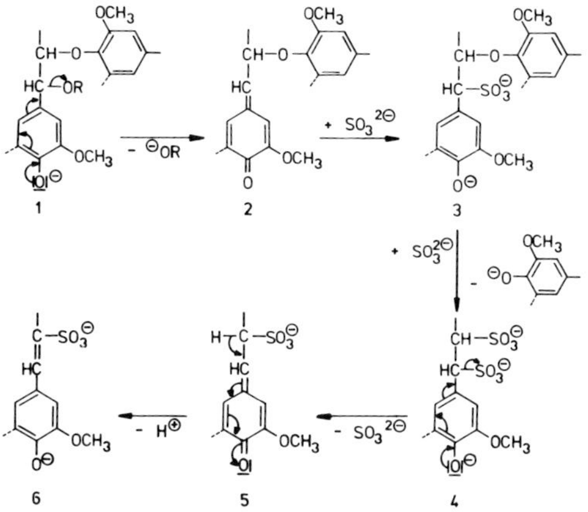

Fig. 7-12.   Reactions of phenolic β-aryl ether and α-ether structures (1) during neutral sulfite pulping (Gierer, 1970). R = H, alkyl, or aryl group. The quinone methide intermediate (2) is sulfonated to structure (3). The negative charge of the α-sulfonic acid group facilitates the nucleophilic attack of the sulfite ion, resulting in β-aryl ether bond cleavage and sulfonation. Structure (4) reacts further with elimination of the sulfonic acid group from α-position to form intermediate (5) which finally after abstraction of proton from β-position is stabilized to a styrene-β-sulfonic acid structure (6). Note that only the free phenolic structures are cleaved, whereas the nonphenolic units remain essentially unaffected.

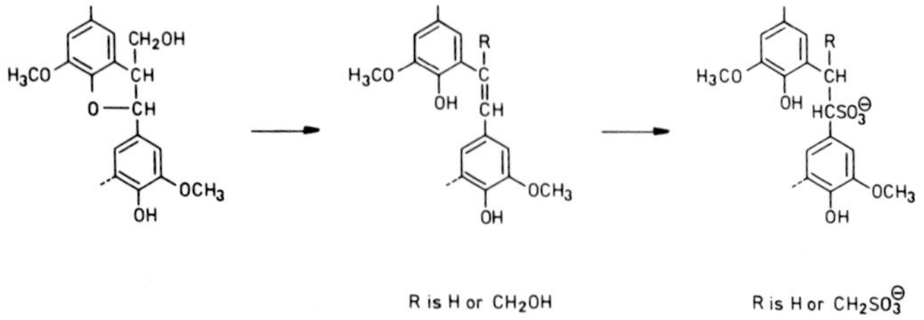
H3CO
CH2OH
CH
O—CH
OCH3
OH
R
C
CH
R is H or CH2OH
R is H or CH2SO3⊖
HCSO3⊖

Fig. 7-13. Reaction of phenyl coumaran structures during neutral sulfite pulping (see Gellerstedt, 1976).

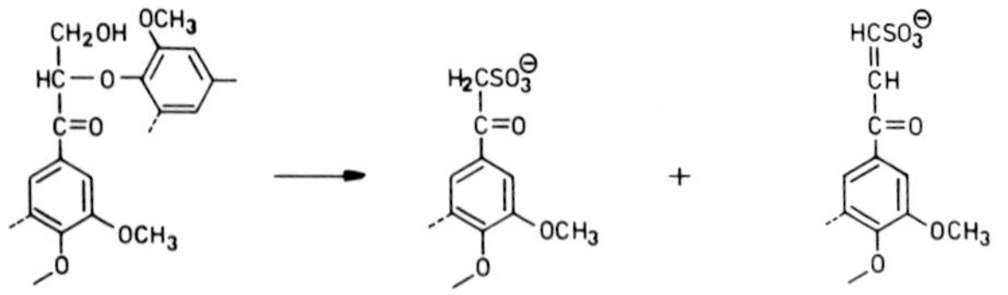
CH2OH
OCH3
HC—O
C=O
H2CSO3⊖
HCSO3⊖
CH
+
O

Fig. 7-14. Cleavage of β-aryl ether bonds in structures containing α-carbonyl groups (see Gellerstedt, 1976). Note that this reaction can take place even in nonphenolic units.

tures and that of 1,4-diarylbutadienes from pinoresinol structures (cf. Fig. 7-13). The conjugated structures can be further sulfonated at their α-carbon atoms. Moreover, the quinone methides can condense with nucleophilic sites of other phenylpropane units or with thiosulfate. The condensation products are similar to those formed during acid sulfite pulping.

Carbonyl groups present in lignin may have a great influence on its reactions with neutral sulfite. For example, α-carbonyl groups can activate the β-aryl ether bonds in nonphenolic units and induce their cleavage (Fig. 7-14). Coniferaldehyde end groups are also extensively sulfonated. Finally, methoxyl groups which are completely stable towards acid sulfite, may, in part, be cleaved during neutral sulfite pulping with the formation of methane sulfonic acid (Fig. 7-15).

Although adequate information is lacking, the reactions of lignin with alkaline sulfite are largely related to those occurring in neutral sulfite and alkali pulping. During alkaline sulfite pulping the β-aryl ether bonds are obviously cleaved also in nonphenolic units, and the condensation reactions have been proposed to be less important compared with those during kraft pulping.

7.2.6 Carbohydrate Reactions

Acid Sulfite Pulping Since glycosidic bonds are easily cleaved by acids, depolymerization of polysaccharides during acid sulfite pulping cannot be

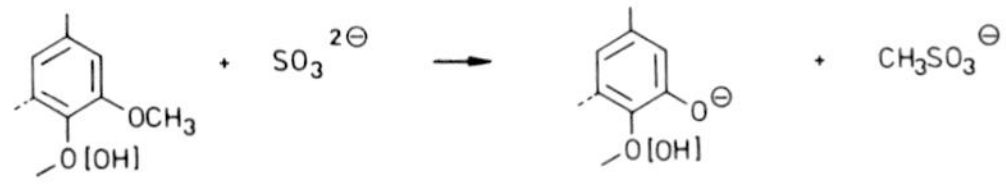

Fig. 7-15. Cleavage of the methyl aryl ether bond with formation of methanesulfonic acid during neutral sulfite pulping (Gierer, 1970).

avoided. Hemicelluloses are attacked more easily than cellulose mainly because of their amorphous state and relatively low degree of polymerization. Most of their glycosidic bonds are also more labile toward acid hydrolysis than the glucosidic bonds in cellulose. When the hydrolysis has proceeded far enough, the degraded hemicelluloses dissolve in the cooking liquor and are gradually hydrolyzed to monosaccharides. Although cellulose is also depolymerized during pulping, practically no losses occur unless the delignification is extended to very low lignin contents (dissolving pulp).

Softwood hemicelluloses contain mainly galactoglucomannans and arabinoglucuronoxylan. Galactose residues are easily hydrolyzed during pulping. The acetyl groups present in galactoglucomannans are also almost completely removed under acidic pulping conditions irrespective of the moderate stability of ester linkages toward acid. Because of these reactions, the major hemicellulose remaining in the final softwood pulp is glucomannan. The arabinonofuranose substituents in arabinoglucuronoxylan are extremely labile toward acid hydrolysis. The remaining glucuronoxylan component in the pulp has been found to contain less glucuronic acid substituents than the original xylan.

Hardwood hemicelluloses are mainly composed of glucuronoxylan which is partly deacetylated during pulping. In addition, a remarkable decrease in the uronic acid content takes place. It is to be noted that the hydrolysis of glycuronide bonds takes place much more slowly than that of the glycosidic bonds between neutral monosaccharide units. The decrease in the uronic acid content is therefore associated with a selective dissolution of xylan during pulping. This means that fragments with high uronic acid contents are readily dissolved whereas chains with fewer acid side chains are preferentially retained in the pulp. Hardwood glucomannan is depolymerized during pulping and most of it is dissolved. In addition to these hemicelluloses, wood contains soluble polysaccharides, such as starch, pectic acid, and arabinogalactan. These components are dissolved at early stages of the cook.

The monosaccharides formed during pulping are not completely stable. About 10–20% of them are oxidized to aldonic acids by hydrogen sulfite ions. Simultaneously, hydrogen sulfite is reduced to thiosulfate:

$$2 \text{ R-CHO} + 2 \text{ HSO}_3^- \rightarrow 2 \text{ R-COOH} + S_2O_3^{2-} + H_2O \qquad (7\text{-}8)$$

$$R = \text{monosaccharide residue}$$

In addition to the aldonic acid formation a minor fraction of the monosaccharides is converted to sugar sulfonic acids.

Generally, no cellulose is lost in the acid sulfite process. The losses of hemicelluloses are higher for hardwood than for softwood. Typical yield values for various wood constituents have been collected in Table 7-3 in which comparative data for the kraft process are also shown.

Two-Stage or Multistage Sulfite Pulping The carbohydrate yield in pulping can be improved considerably by the application of two-stage pulping procedures, e.g., the "Stora method," according to which chips are pretreated with a sodium hydrogen sulfite–sulfite solution under nearly neutral conditions and then subjected to a normal acid sulfite cook. During the first stage, lignin is sulfonated to some degree but remains mainly in the solid phase. Delignification is accomplished in the second, acidic stage. Also pine heartwood can be delignified by the two-stage sulfite method, since in the first stage the most reactive groups of lignin are protected by sulfonation and hence cannot react with pinosylvin and other phenolic compounds in the second stage. Compared to conventional acid sulfite pulping the two-stage method, when applied to softwood, gives 5–7% higher yields (calculated on wood). This increase is largely due to an increased retention of glucomannan (Fig. 7-16). Extensive research has shown that the retention of glucomannan is probably a consequence of the alkaline hydrolysis of its acetyl groups, since there is a close relationship between the acetyl content of the pretreated wood and the pulp (or glucomannan) yield. For cleavage of these ester linkages a treatment with neutral or slightly alkaline solutions at elevated temperature is sufficient. (Note that the "initial" pH of bisulfite

TABLE 7-3. Yields of Various Pulp Constituents after Sulfite and Kraft Pulping of Norway Spruce (*Picea abies*), Scots Pine (*Pinus sylvestris*), and Birch (*Betula verrucosa*)[a]

Constituents	Spruce sulfite		Birch sulfite		Pine kraft		Birch kraft	
Cellulose	41	(41)[b]	40	(40)	35	(39)	34	(40)
Glucomannan	5	(18)	1	(3)	4	(17)	1	(3)
Xylan	4	(8)	5	(30)	5	(8)	16	(30)
Other carbohydrates and various components	—	(4)	—	(4)	—	(5)	—	(4)
Sum of carbohydrates	50	(69)	46	(74)	44	(67)	51	(74)
Lignin	2	(27)	2	(20)	3	(27)	2	(20)
Extractives	0.5	(2)	1	(3)	0.5	(4)	0.5	(3)
Sum of components (total yield)	52	(100)	49	(100)	47	(100)	53	(100)

[a] The figures are calculated as percent of the dry wood.
[b] Figures in parentheses refer to the original wood composition.

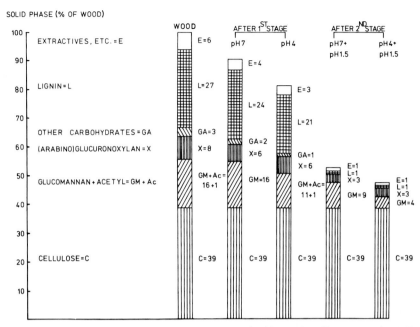

Fig. 7-16. Material balance for a typical two-stage cook of Scots pine (Sjöström et al., 1962).

solutions increases with increasing temperature.) Consequently, the first-stage treatment can result in an almost complete deacetylation. It is believed that the deacetylated glucomannan is more closely hydrogen bonded to the cellulose microfibrils (or partly crystallized) in which state its resistance toward acid hydrolysis is improved. By contrast, two-stage pulping of hardwoods improves the carbohydrate (xylan) yield only moderately in comparison with conventional pulping, possibly because of the presence of glucuronic acid substituents, which might counteract association with cellulose molecules.

Other modifications of multistage sulfite pulping have been proposed and are also used commercially. For example, dissolving pulps with a high cellulose content can be produced by treating the pulp after acid sulfite delignification at a higher pH with sodium carbonate solution in order to remove hemicelluloses. In the case of pine wood, which cannot be delignified with acid sulfite directly, this may result in a three-stage pulping procedure, e.g., the "Sivola method."

Neutral Sulfite Pulping Neutral sulfite pulping is mainly used for the production of high-yield pulps from hardwoods. The process is often named the "NSSC" (Neutral Sulfite Semi-Chemical) process. The wood is delignified only partially and mechanical fiberization is thus required. The product obtained is especially suitable for use as corrugating medium. During neutral sulfite pulping lignin in the solid phase becomes sulfonated to a certain degree, resulting in increased hydrophilicity and swelling. Acetyl groups are almost completely eliminated by base-catalyzed hydrolysis. The selectivity of delignification is relatively good only at the beginning of the delignification. The example given in Fig. 7-17 clearly shows that neutral sulfite pulping is applicable only to the production of high-yield pulps with a high residual lignin content. The mechanical fiberization after pulping causes some material losses, especially in xylan.

7.2.7 Resin Reactions

During sulfite pulping the fatty acid esters are saponified to an extent determined by the conditions. Some of the resin components can also become sulfonated, resulting in increased hydrophilicity and better solubility. However, the partial removal of resin that always occurs during sulfite cooking and subsequent mechanical treatment is mainly associated with the formation of finely dispersed resin particles in stable emulsions. The dissolved lignosulfonic acids act as detergents with respect to the lipophilic resin components.

Acid sulfite pulping causes terpenes, terpenoids, and flavonoids to become partially dehydrogenated. The formation of p-cymene from α-pinene and quercetin from taxifolin are well-known reactions of this type (Fig. 7-18). Due to unsaturation, diterpenoids, including the resin acids, are probably partially polymerized to high molecular weight products causing pitch problems in subsequent pulp handling. Delignification of parenchyma cells with a high content of resin remains incomplete.

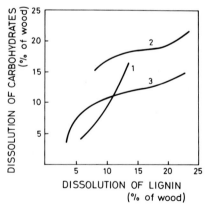

Fig. 7-17. Comparison of delignification selectivities. Neutral sulfite (1), acid sulfite (2), and two-stage sulfite (3) (Sjöström, 1964).

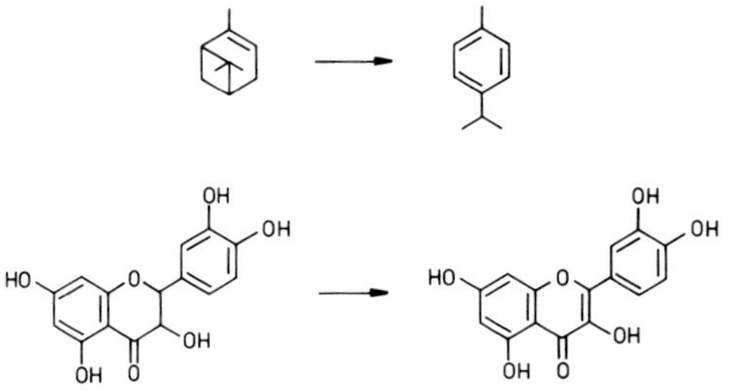

Fig. 7-18. Conversion of α-pinene to p-cymene and taxifolin to quercetin during acid sulfite pulping.

7.2.8 Side Reactions

A considerable part of the hydrogen sulfite ions is consumed in reactions other than the sulfonation of lignin. In the absence of wood, sulfur dioxide solutions are decomposed at elevated temperatures according to the equation:

$$4\ HSO_3^- \rightarrow S_2O_3^{2-} + 2\ SO_4^{2-} + 2\ H^+ + H_2O \qquad (7\text{-}9)$$

This disproportionation reaction follows complex kinetics involving formation of polythionates ($S_3O_6^{2-}$, $S_4O_6^{2-}$, and $S_5O_6^{2-}$) as intermediate products. The initial and rather slow decomposition during the induction period is later accelerated by thiosulfate, which functions as an autocatalyst. After a critical thiosulfate concentration has been reached, sulfur is precipitated, and the acidity increases rapidly.

Another mechanism giving rise to thiosulfate formation is the reduction of hydrogen sulfite by wood components. Sugars play an important role in this reduction (see p. 117). Similarly, α-pinene is oxidized to p-cymene, formic acid to carbon dioxide, and taxifolin present in Douglas fir heartwood to quercetin. Sulfite cooking liquors, however, contain less thiosulfate than might be expected on the basis of these side reactions (Table 7-4). This discrepancy can be attributed to continuous consumption of thiosulfate in reactions with lignin, producing thioether cross-links (see p. 115). This type of sulfur, termed "organic excess sulfur," may account for 5–10% of the total organically bound sulfur. The major portion, 80–90% of the sulfur, exists in the form of sulfonate groups in the lignin, although minor amounts of sulfite are also consumed in the formation of carbohydrate sulfonic acids.

7.2.9 Composition of Sulfite Spent Liquors

In addition to lignosulfonates and hemicelluloses and their degradation products, sulfite spent liquors contain small amounts of uronic acids, methyl

TABLE 7-4. Origins of Thiosulfate Formation during Sulfite Pulping[a]

Cause	$S_2O_3^{2-}$ (g/liter)	Sulfur loss (kg/ton of pulp)
Aldonic and sugar sulfonic acid formation (5% sugar composition, wood basis)	4.0	23
Carbonic acid formation (0.3% formic acid, wood basis)	0.9	5
Cymene formation (0.5 kg ptp)	0.02	—
Sulfate formation (2–3 g/liter)	1.5	9
Total thiosulfate formation	6.5	37
Found as thiosulfate	0.5–2	
Found as polythionate (expressed as thiosulfate)	0.5–1.5	
Total thiosulfate found	1–3.5	5–20

[a] From Rydholm (1965).

glyoxal, formaldehyde, methyl alcohol, furfural, etc. Most of the remaining sulfur dioxide in the liquor is directly titratable, but some of it is liberated slowly under titration or after certain treatments. The α-hydroxysulfonic acids derived from carbonyl compounds are responsible for this "loosely combined sulfur dioxide" (cf. Figs. 2-29 and 7-7). In the liquors from acid sulfite cooking most of the carbohydrates are present in the form of monosaccharides (Table 7-5, cf. also Table 10-1). After bisulfite and neutral sulfite pulping, however, a large portion of the sugars remains as oligo- and polysaccharides. Characteristic of the spent liquors from neutral sulfite cooking of hardwood is the high proportion of acetic acid in comparison with the other organic constituents present.

TABLE 7-5. Typical Composition of the Sulfite Spent Liquor Resulting from the Acid Sulfite Pulping of Norway Spruce

Component	Content (% of dry solids)	Composition (% of carbohydrates)
Lignosulfonates	55	
Carbohydrates	28	
Arabinose		4
Xylose		22
Mannose		43
Galactose		17
Glucose		14
Aldonic acids	5	
Acetic acid	4	
Extractives	4	
Other compounds	4	

7.2.10   Recovery of Sulfite Cooking Chemicals

In addition to the organic solids resulting from the degradation and dissolution of the wood constituents the spent cooking liquors contain the inorganic pulping chemicals, which for the most part have been changed during the pulping reactions. A variety of useful products can be produced from this organic source (see Section 10.2), but most of the solids in the spent liquor are still burned with generation of heat. Beyond the heat economy aspects, the combustion of the organic substance is today necessary from the pollution point of view.

Because calcium was long used almost exclusively in the sulfite process as the base, no need existed for the recovery of this comparatively cheap chemical. Interest was therefore only directed toward the relatively simple recovery of excess sulfur dioxide. However, the introduction of the more expensive soluble bases, together with more stringent environmental requirements, stimulated the development of methods for the recovery of both heat and inorganic chemicals (base and sulfur).

The concentration of solids in the sulfite spent liquors in the digester after cooking is 11–17%, dropping to 10–15% after pulp washing. For proper ignition and burning, the spent liquors are in most cases concentrated in multiple-effect evaporators to a solids content of 50–65%. Because volatile components, such as acetic and formic acids and furfural, are transferred to the steam condensates upon evaporation, the handling of these dilute liquors requires specific procedures to avoid water pollution. After combustion the resulting ash, consisting of a mixture of calcium sulfate and calcium oxide in roughly equal amounts, is not converted for reuse in pulping; only the dust is collected to reduce air pollution. The stack gases resulting from the combustion of calcium-based spent liquors are especially problematic, because of their high content of sulfur dioxide which can be neither economically recovered nor eliminated.

A furnace similar to the Tomlinson kraft recovery furnace is used for the combustion of magnesium-based sulfite spent liquors. In this case, however, no smelt is obtained; instead the base is completely recovered as magnesium oxide in dust collectors; the sulfur escapes as sulfur dioxide and is absorbed from the combustion gases in scrubber towers. However, because magnesium hydroxide has a very low solubility in water, a complete recovery of sulfur dioxide meets difficulties.

Ammonium-based sulfite spent liquors can be burned in the same type of furnace as the calcium-based liquors. However, during combustion the base is decomposed to form nitrogen and water and the problems with fly ash are thus eliminated. All sulfur escapes to the combustion gases as sulfur dioxide which can be partly absorbed in an ammonia solution.

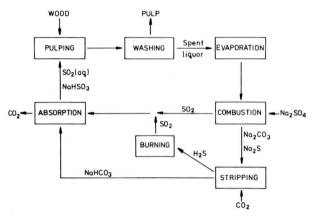

Fig. 7-19. Recovery and conversion cycles for sulfite cooking chemicals.

Sodium-based spent liquors both from the acid and neutral sulfite processes can be burned in a kraft type furnace. A smelt is obtained, consisting of sodium sulfide and sodium carbonate. The sulfur-to-sodium molar ratio is about 1:1 for sulfite spent liquors instead of around 0.15:1 in kraft black liquors, which means that a considerable part of the sulfur escapes as sulfur dioxide. Sodium carbonate from the recovery cycle is suitable for the absorption of the sulfur dioxide, although special problems are encountered because the sulfur dioxide concentration of these gases is low and oxidation to sulfur trioxide must be avoided. However, the greatest problem is the complete removal of sulfide from the smelt and its conversion to pure cooking chemicals. Especially for acid sulfite pulping, the cooking chemicals must be very pure, since other sulfur compounds, especially thiosulfate, are detrimental for pulping. Several recovery systems based on the use of a conventional kraft recovery furnace have been developed. In another and a simpler system, which, however, requires a modified furnace construction, all sulfur is first converted to hydrogen sulfide together with partial or complete gasification of the organic constituents. Sodium is thus recovered as pure carbonate. Hydrogen sulfide is finally converted to sulfur dioxide and absorbed. The recovery and conversion cycles for sulfite cooking chemicals are illustrated in Fig. 7-19.

7.3 Kraft Pulping

7.3.1 Cooking Chemicals and Equilibria

Sulfate or kraft pulping is performed with a solution composed of sodium hydroxide and sodium sulfide, named "white liquor." According to the terminology the following definitions are used, where all the chemicals are calculated as sodium equivalents and expressed as weight of NaOH or Na_2O.

Total alkali	All sodium salts
Titratable alkali	$NaOH + Na_2S + Na_2CO_3$
Active alkali	$NaOH + Na_2S$
Effective alkali	$NaOH + \frac{1}{2} Na_2S$
Causticizing efficiency	$100 \dfrac{NaOH}{NaOH + Na_2CO_3}$ %
Sulfidity	$100 \dfrac{Na_2S}{NaOH + Na_2S}$ %
Degree of reduction	$100 \dfrac{Na_2S}{Na_2S + Na_2SO_4}$ %

In modern pulping chemistry weight units of NaOH are often replaced by molar units, e.g., moles of effective alkali per liter of solution or kilogram of wood. Table 7-6 shows typical conditions for kraft pulping. The charge of effective alkali (NaOH) applied is usually 4–5 moles or 16–20% of wood.

The following equilibria are involved in the aqueous solutions containing sodium sulfide and sodium hydroxide:

$$S^{2-} + H_2O \rightleftarrows HS^- + HO^- \tag{7-10}$$

$$HS^- + H_2O \rightleftarrows H_2S + HO^- \tag{7-11}$$

The equilibrium constants for these reactions are:

$$K_1 = [HS^-][HO^-]/[S^{2-}] \tag{7-12}$$

$$K_2 = [H_2S][HO^-]/[HS^-] \tag{7-13}$$

Since $K_1 \sim 10$ and $K_2 \sim 10^{-7}$, the equilibrium in equation 7-10 strongly favors the presence of hydrosulfide ions and for all practical purposes, sulfide ions can be considered to be absent (Fig. 7-20). The concentration of

TABLE 7-6.   Alkaline Pulping Methods and Conditions

Method	pH range	"Base"	Active reagents	Max. temp. (°C)	Time at max. temp. (hr)	Softwood pulp yield (%)
Alkali (soda)	13–14	Na^+	HO^-	155–175	2–5	50–70[a]
Kraft	13–14	Na^+	HS^-, HO^-	155–175	1–3	45–55
Prehydrolysis–kraft						35–40
Prehydrolysis stage	3–4		H^+	160–175	0.5–3	
Kraft stage	13–14	Na^+	HS^-, HO^-	155–175	1–3	

[a] Hardwood.

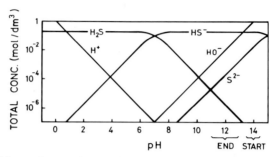

Fig. 7-20. Equilibrium diagram showing the composition of the kraft white liquor at different pH values. Concentration, 0.2 mole/liter; temp., 25°C. Based on equilibrium constants corresponding to $K_1 = 10$ and $K_2 = 10^{-7}$.

hydrogen sulfide becomes significant below pH 8 and needs to be considered only in modified kraft pulping processes involving pretreatment at low pH (see Section 7.3.6).

7.3.2 Impregnation

In the kraft process thorough impregnation of the chips with cooking chemicals is not as critical as in acid sulfite pulping. The diffusion of chemicals in liquid-saturated wood is controlled by the total cross-sectional area of all the capillaries. In moderately alkaline solutions (pH < 12.5) the *effective capillary cross sectional area* (ECCSA), which is the area of paths available for diffusion, is higher in the longitudinal direction than in the radial and tangential directions (Fig. 7-21). However, because of swelling caused by alkali at pH values above 13, the ECCSA is increased in tangential and radial

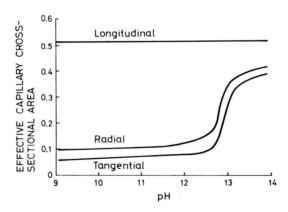

Fig. 7-21. Effective capillary cross-sectional area (ECCSA) of aspen wood as a function of pH (Stone, 1957). © 1957. TAPPI. Reprinted from *Tappi* **40**(7), p. 54, with permission.

directions, approaching the same permeability as in the longitudinal direction.

7.3.3 General Aspects of Delignification

The consumption of effective alkali in a kraft cook corresponds to about 150 kilogram sodium hydroxide per ton of wood. As a result of the alkaline degradation of polysaccharides, about 1.6 equivalents of acids are formed for every monosaccharide unit peeled from the chain. Of the charged alkali, 60–70% is required for the neutralization of these hydroxy acids, while the rest is consumed to neutralize uronic and acetic acids (about 10% of alkali) and degradation products of lignin (25–30% of alkali).

Hydrosulfide ions react with lignin, but most of the sulfur-containing lignin products are decomposed during the later stages of the cook with formation of elemental sulfur which combines with hydrosulfide ions to form polysulfide. However, kraft lignin still contains 2–3% of sulfur corresponding to 20–30% of the charge.

Figure 7-22 illustrates the dissolution of carbohydrates and lignin during kraft pulping. The carbohydrates are attacked already at a comparatively low temperature. This means that the acetyl groups are completely removed and the peeling process is terminated long before the maximum cooking temperature has been attained. The reactivity of the polysaccharides varies depending on their accessibility as well as on their structure. Because of its crystalline nature and high degree of polymerization, cellulose suffers less losses than the hemicelluloses.

The dissolution of lignin can be divided into three phases (Fig. 7-23). The

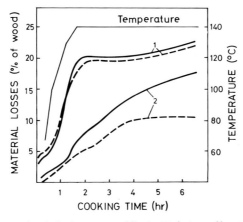

Fig. 7-22. Dissolution of carbohydrates (1) and lignin (2) during sulfate (———) and soda (- - -) pulping of Norway spruce (see Enkvist et al., 1957). Note that the cooking temperature is exceptionally low (140°C), resulting in much less dissolution than in normal pulping.

initial phase of delignification takes place at temperatures below 140°C and is controlled by diffusion. Above 140°C, the rate of delignification becomes controlled by chemical reactions and accelerates steadily with increasing temperature. The rate of lignin dissolution remains high during this "bulk delignification" phase, until about 90% of the lignin has been removed. The final slow phase is termed "residual delignification" and can be regulated to some degree by varying the alkali charge and the cooking temperature.

The kinetics of the delignification are of importance especially when considering the control of the pulping process. Since kraft pulping follows simpler kinetics than the sulfite processes, more applications have been adopted for this case. Because of the heterogenity of the system, however, pulping reactions are complicated and can therefore not be treated in the same fashion as homogeneous reactions in solution.

The overall rate of the bulk delignification in kraft pulping, during which the variations in hydroxyl and hydrosulfide ion concentrations are moderate, follows pseudo-first-order kinetics, approximately in conformity with the following equation:

$$- \frac{dL}{dt} = kL \tag{7-14}$$

where L is the lignin content of wood residue at time t and k the rate constant.

Based on experimental data of k at varying temperatures, the value of the activation energy E_a can be calculated from the Arrhenius equation:

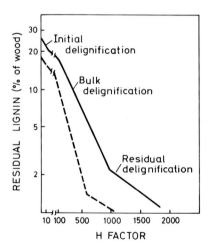

Fig. 7-23. Removal of lignin during kraft pulping of pine (——) and birch (- - -) as a function of the H factor (Kleppe, 1970). © 1970. TAPPI. Reprinted from *Tappi* **53**(10), p. 37, with permission.

$$\ln k = \ln A - \frac{E_a}{RT} \qquad (7\text{-}15)$$

where T is the absolute temperature (Kelvin), R the gas constant, and A a further constant including the frequency factor. The activation energy E_a for kraft delignification of softwood has been determined to be 130–150 kJ/mole (31–36 kcal/mole) (bulk phase) and about 50 and 120 kJ/mole (12 and 27 kcal/mole) for the initial and final phases, respectively.

According to a simplified system the net effect of both cooking time and temperature can be expressed by means of a single variable. In this system the rate at 100°C is chosen as unity and rates at all other temperatures are related to this standard. When using a value of 134 kJ/mole (32 kcal/mole) for E_a the rates at any other temperatures can then be expressed by the following equation:

$$\ln \text{(relative reaction rate)} = 43.2 - \frac{16,113}{T} \qquad (7\text{-}16)$$

When the relative reaction rate is plotted against cooking time, the area under the resulting curve is called H factor:

$$H = \int_0^t e43.20 - \frac{16,113}{T} \, dt \qquad (7\text{-}17)$$

A normal heating period contributes to the H factor by 150–200, and 1500–2000 are needed for a complete kraft cook. Within the bulk delignification phase the relative reaction rate is doubled when the temperature is increased by about 8°C. If the hydroxide and hydrosulfide ion concentrations vary in a reproducible manner the H factor will predict the degree of delignification with sufficient precision.

7.3.4 Lignin Reactions

As in sulfite pulping, fragmentation of lignin depends on the cleavage of ether linkages, whereas the carbon-to-carbon linkages are essentially stable. The presence of hydrosulfide ions greatly facilitates delignification because of their strong nucleophilicity in comparison with hydroxyl ions, which, instead, are strongly basic. Cleavage of ether linkages, promoted both by hydroxyl and hydrosulfide ions, results also in increasing hydrophilicity of lignin because of the liberation of phenolic hydroxyl groups. The degraded lignin is dissolved in the cooking liquor as sodium phenolates. Studies with model substances representing various structural units in lignin have largely clarified the delignification reactions in kraft pulping.

Etherified Phenolic Structures Containing β-Aryl Ether Bonds   In etherified *p*-phenolic structures the β-aryl ether linkage is cleaved by hydroxide

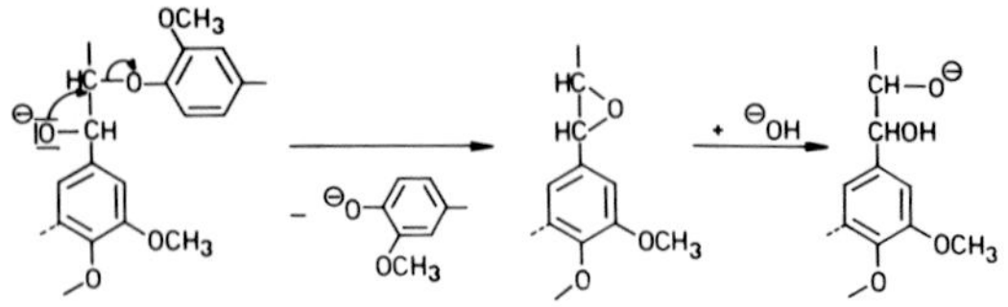

Fig. 7-24.   Cleavage of β-aryl ether bonds in nonphenolic phenylpropane units during soda pulping (Gierer, 1970).

ions according to the mechanism shown in Fig. 7-24. The reaction proceeds via an oxirane intermediate which is subsequently opened with formation of an α,β-glycol structure. This reaction promotes efficient delignification by fragmenting the lignin and by generating new free phenolic hydroxyl groups.

Free Phenolic Structures Containing β-Aryl Ether Bonds   The first step of the reaction involves the formation of a quinone methide from the phenolate anion by the elimination of a hydroxide, alkoxide, or phenoxide ion from the α-carbon (Fig. 7-25). The subsequent course of reactions depends on whether hydrosulfide ions are present or not. In the latter case (soda pulping), the dominant reaction is the elimination of the hydroxymethyl group from the quinone methide with formation of formaldehyde and a styryl aryl ether structure *without* cleavage of the β-ether bond (Fig. 7-26). When hydrosulfide ions are present (strong nucleophiles) they react with the

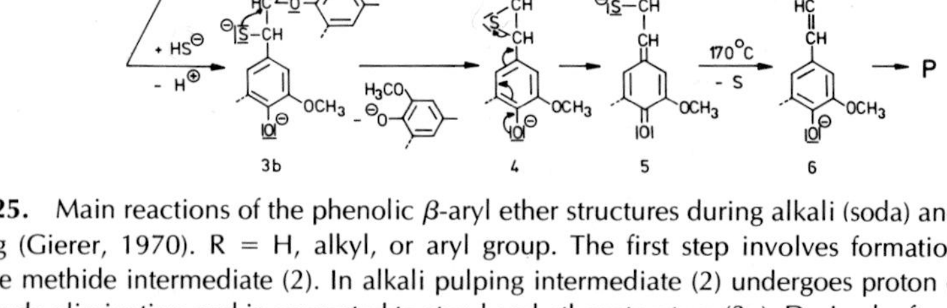

Fig. 7-25.   Main reactions of the phenolic β-aryl ether structures during alkali (soda) and kraft pulping (Gierer, 1970). R = H, alkyl, or aryl group. The first step involves formation of a quinone methide intermediate (2). In alkali pulping intermediate (2) undergoes proton or formaldehyde elimination and is converted to styryl aryl ether structure (3a). During kraft pulping intermediate (2) is instead attacked by the nucleophilic hydrosulfide ions with formation of a thiirane structure (4) and simultaneous cleavage of the β-aryl ether bond. Intermediate (5) reacts further either *via* a 1,4-dithiane dimer or directly to compounds of styrene type (6) and to complicated polymeric products (P). During these reactions most of the organically bound sulfur is eliminated as elemental sulfur.

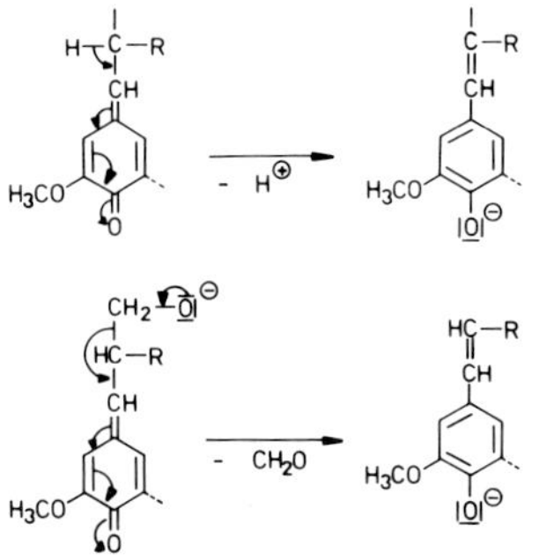

Fig. 7-26. Elimination of proton and formaldehyde from the quinone methide intermediate during alkali pulping (Gierer, 1970).

quinone methide to form a thiol derivative which is converted to a thiirane structure with simultaneous cleavage of the β-ether bond. The thiirane can be dimerized to a dithiane structure, but this as well as other sulfur-containing intermediates are decomposed, forming elemental sulfur and unsaturated side-chain structures. Competing reaction paths are possible, giving rise to other minor degradation products, such as guaiacol.

Structures Containing α-Ether Bonds   The α-ether bonds in phenolic phenylcoumaran (Fig. 7-27) and pinoresinol structures are readily cleaved by hydroxide ions, usually followed by the release of formaldehyde. Only in the case of open α-aryl ether structures does this reaction result in the fragmentation of lignin. In contrast, the α-ether bonds are stable in all etherified structures.

Methoxyl Groups   Lignin is partially demethylated by the action of hydrosulfide ions forming methyl mercaptan which is convertible to dimethyl sulfide by reaction with another methoxyl group. In the presence of oxygen, methyl mercaptan can be oxidized further to dimethyl disulfide (Fig. 7-28). Because the hydroxide ions are less strong nucleophiles than hydrosulfide ions, only small amounts of methanol are formed. Methyl mercaptan and

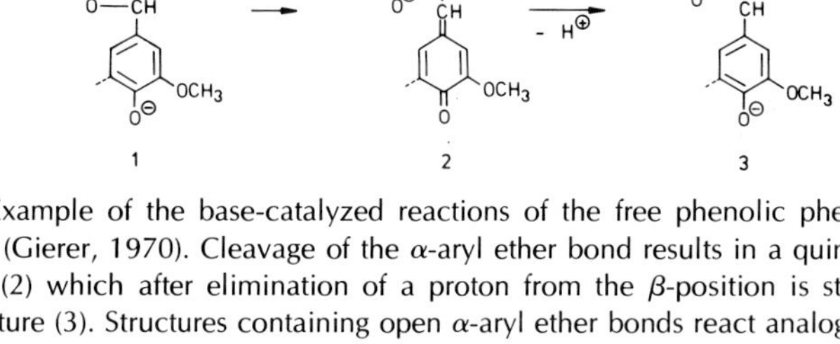

Fig. 7-27. Example of the base-catalyzed reactions of the free phenolic phenylcoumaran structures (1) (Gierer, 1970). Cleavage of the α-aryl ether bond results in a quinone methide intermediate (2) which after elimination of a proton from the β-position is stabilized to a stilbene structure (3). Structures containing open α-aryl ether bonds react analogously.

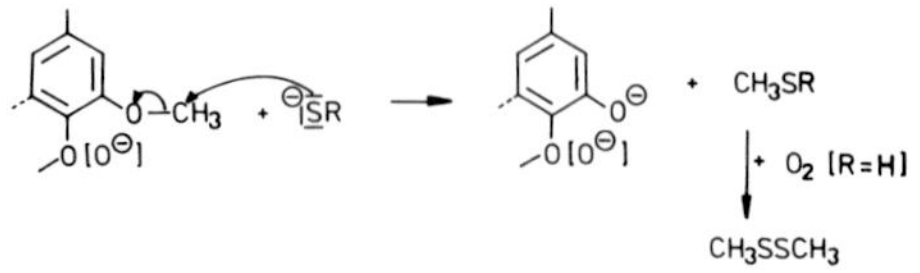

Fig. 7-28. Cleavage of methyl aryl ether bonds with simultaneous formation of methyl mercaptan (CH_3SH), dimethyl sulfide (CH_3SCH_3), and dimethyl disulfide (CH_3SSCH_3) during kraft pulping (Gierer, 1970). R = H or methyl group.

dimethyl sulfide are highly volatile and extremely malodorous, causing an air pollution problem that is difficult to master.

Condensation Reactions   A variety of condensation reactions are known to occur in alkaline pulping. Since carbon-to-carbon linkages are formed between lignin entities, it has been proposed that as a result of condensation reactions, lignin dissolution is retarded, particularly during the terminal phases of kraft pulping.

It has been suggested that the major part of condensation processes occurs at the unoccupied C-5 position of phenolic units. Thus, in isolated MWL preparations about half of the C-5 positions are unsubstituted, while in isolated kraft lignins only about one third of these positions remain free. The syringyl units of hardwood lignins cannot, of course, undergo condensation reactions of this type.

Figure 7-29 shows some examples of postulated condensation reactions

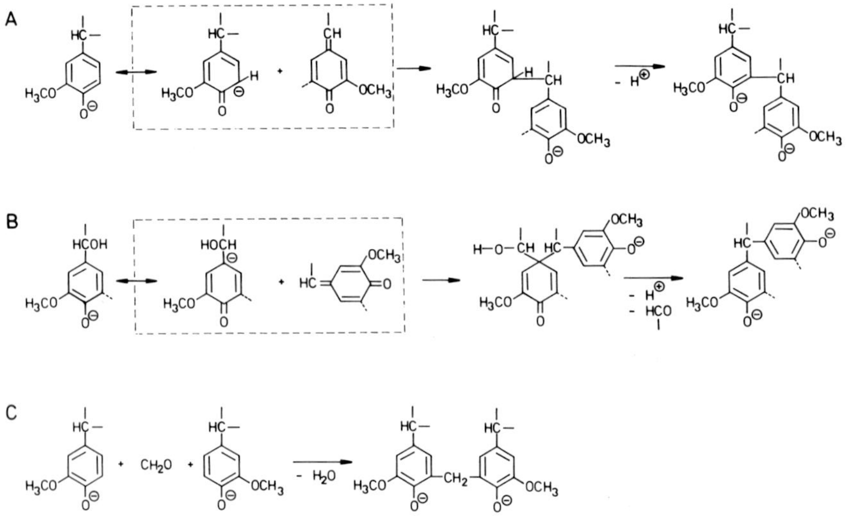

Fig. 7-29. Examples of condensation reactions during alkali and kraft pulping (Gierer, 1970).

forming diarylmethane structures of three different types. In the first case (A) a phenolate adds to a quinone methide structure, forming an α-5 linkage. The second case (B) illustrates a similar condensation between the 1- and α-carbons with simultaneous removal of the propane side chain. The third reaction (C) involves formaldehyde released from the γ-carbinol groups (see Fig. 7-26) and also leads ultimately to a diarylmethane structure.

Formation of Chromophores   During kraft delignification the color of the chips darkens gradually until a yield level of 60–70% is reached (light absorption coefficient of ca. 40 m^2/kg instead of ca. 5–10 m^2/kg for the original wood at 457 nm). Further delignification brings about a modest brightening of the pulp. The specific light absorption coefficient of the residual lignin, however, increases continuously, reaching ca. 500 m^2/kg at the end of the pulping. For comparison, the corresponding value for wood lignin is 20–40 m^2/kg.

The color of unbleached pulps is caused by certain unsaturated structures (chromophores). In addition, leucochromophores, which can be converted into chromophores by air oxidation may be present in the pulp. Most of the chromophores are presumed to be derived from lignin (Fig. 7-30) although some chromophoric groups can also be introduced into the polysaccharides, for example, carbonyl groups.

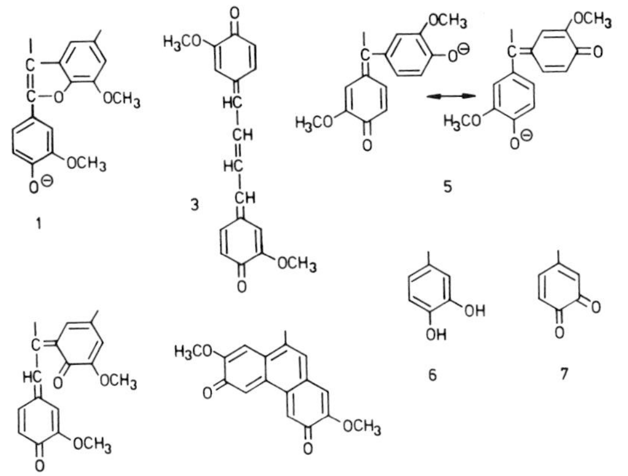

Fig. 7-30. Examples of proposed leucochromophoric and chromophoric structures. Aryl-coumarones (1) and stilbene quinones (2) are thought to be formed from stilbenes after oxidation. Butadiene quinones (3) could arise from oxidation of hydroxyarylbutadienes being formed from phenolic pinoresinol structures during kraft or neutral sulfite pulping. Cyclization may yield intermediates which are further oxidized to cyclic diones (4). A resonance-stabilized structure (5) results from the corresponding condensation product formed during pulping. o-Quinoid structures (7) are oxidation products of catechols (6) formed during alkaline or neutral pulping processes.

7.3.5 The Reactions of the Polysaccharides

Because of the alkaline degradation of polysaccharides kraft pulping results in considerable carbohydrate losses. The acetyl groups are hydrolyzed at the very beginning of the kraft cook (from hardwood xylan and softwood galactoglucomannans). In the earlier stages of cooking the polysaccharide chains are peeled directly from the reducing end groups present (primary peeling). As a result of the alkaline hydrolysis of glycosidic bonds, occurring

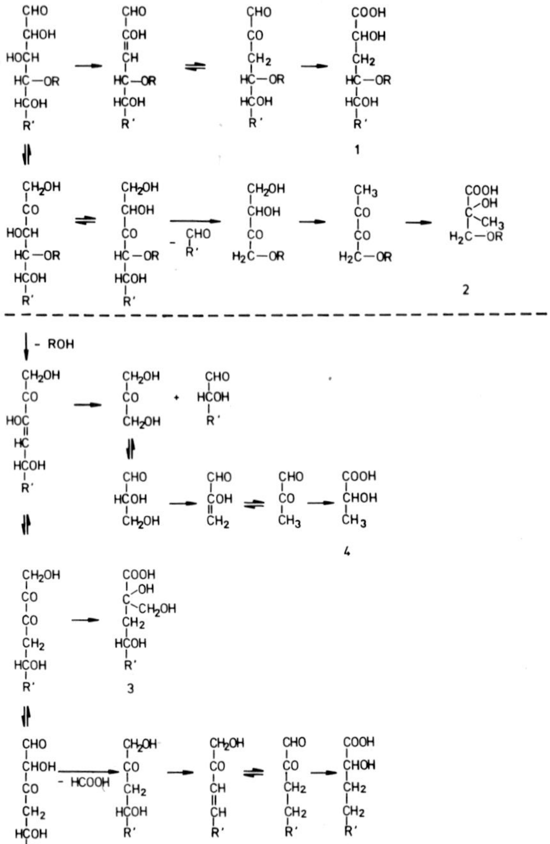

Fig. 7-31. Peeling and stopping reactions of polysaccharides (Sjöström, 1977). R = polysaccharide chain and R' = CH₂OH (cellulose and glucomannans) or H (xylan). Cellulose and glucomannans (R' = CH₂OH): (1) 3-Deoxyhexonic acid end groups (metasaccharinic acid), (2) 2-C-methylglyceric acid end groups, (3) 3-deoxy-2-C-hydroxymethylpentonic acid (glucoisosaccharinic acid), (4) 2-hydroxypropanoic acid (lactic acid), and (5) 3,4-dideoxypentonic acid (2,5-dihydroxypentanoic acid). Xylan (R' = H): (3) 3-Deoxy-2-C-hydroxymethyltetronic acid (xyloisosaccharinic acid), (4) 2-hydroxypropanoic acid (lactic acid), and (5) 2-hydroxybutanoic acid. © 1977. TAPPI. Reprinted from *Tappi* **60**(9), p. 152, with permission.

at high temperatures, new end groups are formed, giving rise to additional degradation (secondary peeling) (cf. Section 2.5.5). As a consequence, the yield of cellulose is always reduced in kraft pulping, although to a lesser extent than that of the hemicelluloses which are degraded more extensively due to their low degree of polymerization and amorphous state. The peeling reaction is finally interrupted because the competing "stopping reaction" converts the reducing end group to a stable carboxylic acid group.

Mechanism of the Peeling Reaction The carbohydrate material lost in peeling is converted to various hydroxy acids. In addition, formic and acetic acid and small amounts of dicarboxylic acids, are formed as well. Fig. 7-31 (lower portion) shows a simplified reaction scheme illustrating the mechanism of formation of the main degradation products. The rearrangement of a reducing end group to a 2-keto intermediate is followed by β-alkoxy elimination. The cleaved monosaccharide unit is rearranged into a 2,3-diulose structure from which either glucoisosaccharinic acid (cellulose and glucomannans) or xyloisosaccharinic acid (xylan) is formed via a benzilic acid rearrangement. The diulose structure can also be cleaved by reversed aldol condensation to glyceraldehyde, which is then converted via methylglyoxal to lactic acid. Finally, a probable route for the formation of 2,5-dihydroxypentanoic and 2-hydroxybutanoic acids proceeds via formic acid elimination from the 3-keto intermediate, followed by a benzilic acid rearrangement.

Figure 7-32 illustrates the glucomannan and xylan losses during kraft pulping of pine wood. As can be seen, an appreciable portion of the lost xylan is actually not degraded but dissolves in the cooking liquor as a polysaccharide. The amount of dissolved xylan reaches a maximum around the midpoint of the delignification process.

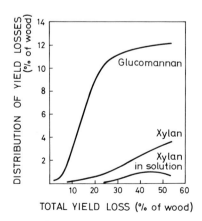

Fig. 7-32. Hemicellulose losses during sulfate pulping of pine wood (Sjöström, 1977). © 1977. TAPPI. Reprinted from *Tappi* **60**(9), p. 152, with permission.

Most of the carbohydrate losses take place during the heating-up period; hence, alkaline hydrolysis of glycosidic bonds does not appreciably contribute to the initial loss. As can be seen from the data in Table 7-3, more than 30% of the wood polysaccharides is lost during kraft pulping. The yield loss is especially high for glucomannan (which is present as galacto-glucomannans in the original wood), but cellulose losses occur as well. The relatively high carbohydrate yield for hardwood kraft pulp compared with hardwood sulfite pulp and softwood kraft pulp depends on the fact that hardwood contains xylan as the dominant hemicellulose component, and xylan is comparatively resistant under alkaline pulping conditions.

Mechanism of End Group Stabilization   The main stopping reaction routes are presented in the upper portion of Fig. 7-31. The dominant route is initiated by a β-hydroxy elimination directly from the aldehydic end groups. The resulting dicarbonyl intermediate is converted to a metasaccharinic acid end group via a benzilic acid rearrangement. The end groups can also rearrange to a 3-keto intermediate, which then loses the 5- and 6-carbons as glycolaldehyde in a reversed aldol condensation. The rest of the end group undergoes β-hydroxy elimination followed by a benzilic acid rearrangement. As a result, a 2-C-methylglyceric acid end group is formed. In addition to the metasaccharinic and 2-C-methylglyceric acid end groups, small amounts of 2-C-methylribonic (glucosaccharinic) and aldonic acid end groups have been found to be present. The presence of aldonic acid end groups indicates that some oxidative reactions also occur. In the case of pulping with alkali alone (soda process), the oxidation might depend on the presence of dissolved oxygen, whereas the polysulfides generated during kraft pulping can function as oxidants.

Softwood xylan is partially substituted with arabinose at the C-3 position of the xylose units. During the course of peeling, arabinose is easily eliminated from the chain under simultaneous formation of a metasaccharinic acid end group (β-alkoxy elimination) which stabilizes the chain against further alkaline peeling. However, because of its relatively low content in softwood, xylan makes only a small contribution to the total carbohydrate yield in pulping. Much more important is the behavior of hardwood xylan during kraft pulping. In this connection, the detailed structure of the xylan chain is of interest (Fig. 3-16). The terminal xylose unit is rapidly cleaved from the xylan chain, but the remaining galacturonic acid end group is stable against further peeling. Its stability is not permanent, however (Fig. 7-33). After hydroxy elimination at C-3, a 2-enuronic acid group is formed which is decomposed at higher temperatures (> 100°C) after isomerization of the double bond to the C-3–C-4 position. The remaining terminal rhamnose unit is eliminated very easily because its C-3 position is bound to the following xylose unit.

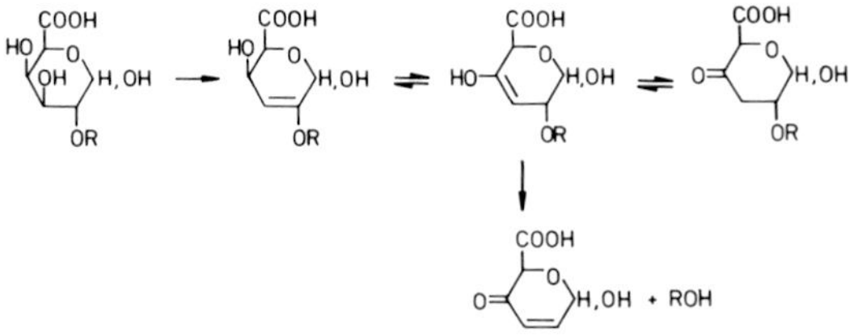

Fig. 7-33. Alkaline degradation of 2-O-(α-L-rhamnopyranosyl)-D-galacturonic acid (Johansson and Samuelson, 1977). R represents the rhamnose group in the xylan chain.

The 4-O-methylglucuronic acid groups prevent the peeling of xylan chains at lower temperatures (< 100°C) but they offer only a partial protection at higher temperatures. Since the 4-O-methylglucuronic acid groups are bound to the C-2 position in the xylose units, no conversion of this carbon atom to a carbonyl group can take place. Instead, HO-3 is eliminated directly (β-hydroxy elimination).

Significance of Uronic Acid Groups The uronic acid content is much lower in the final kraft pulp than in the original wood. In the glucuronoxylan remaining in the pulp it corresponds to a molar ratio of roughly 1:25 (glucuronic acid/xylose), whereas this ratio for native xylan is about 1:5 (softwood) and 1:10 (hardwood). Since xylan is not evenly substituted by the uronic acid groups it is probable that the fractions of high uronic acid content are preferably dissolved, resulting in a lower uronic acid content of the pulp. Another reason for the decrease in the uronic acid content is the cleavage of these groups from the xylan chain, since the pyranosyluronic acid linkages are more sensitive to alkaline hydrolysis than are the corresponding glycosyl linkages. An additional reaction possibly proceeds via CH_3O elimination at C-4 (β-position to the carboxyl group) followed by H-5 elimination (Fig. 7-34).

Other Reactions In addition to the direct stabilizing effect of the uronic acid groups, the relatively high yield of hardwood kraft pulp is also due to the readsorption of xylan on the fibers (Fig. 7-35). After kraft pulping of softwood, the glucomannan remaining in the pulp still contains traces of

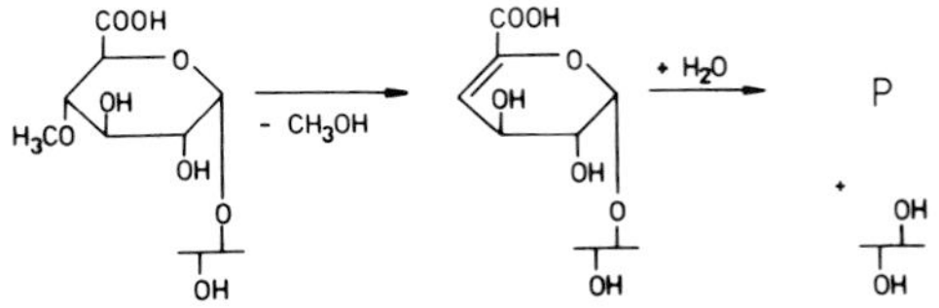

Fig. 7-34. Loss of 4-O-methylglucuronic acid groups (Johansson and Samuelson, 1977). P denotes fragmentation products formed.

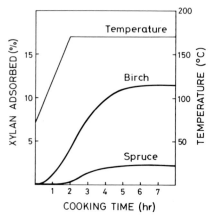

Fig. 7-35. Adsorption of xylan on cotton fibers present in the digester during kraft pulping (Yllner and Enström, 1956).

galactose residues, and the xylan has some arabinose residues contrary to sulfite pulping during which these moieties are cleaved completely.

7.3.6 Stabilization of Polysaccharides against Alkaline Degradation

The primary peeling of polysaccharides by alkali can be avoided by the elimination of the aldehyde functions from the end groups. The reduction of these groups to alcohols by sodium borohydride inhibits primary peeling, and the carbohydrate yield is thus increased considerably. The end groups can also be stabilized by oxidizing them to carboxyl groups or by conversion to other stable derivatives.

Of the stabilization methods, the polysulfide pulping process is of practical importance. The influence of polysulfides is based on a specific oxidation of the end groups to carboxyl groups via glucosone intermediates (cf. Section 8.1.3). Polysulfides can be prepared by catalytic oxidation of sulfide in the white liquor or by adding elemental sulfur into the kraft cooking liquor:

$$n \, S^{2-} + \frac{n-1}{2} \, O_2 + (n-1)H_2O \rightarrow S_n^{2-} + (2n-2)HO^- \qquad (7\text{-}18)$$

$$S^{2-} + n \, S \rightarrow S_{n+1}^{2-} \qquad (7\text{-}19)$$

In the latter method an excess of sulfide is created because of the added sulfur. This must be regenerated to elemental sulfur in order to avoid high sulfidities.

Of the reducing methods, the pretreatment of wood chips with hydrogen sulfide (140°C, pH ~ 7) might be technically feasible. During such a treatment the aldehyde end groups are reduced to thioalditols according to the following equation:

$$R-CHO \underset{}{\overset{H^+, HS^-}{\rightleftharpoons}} R-CH(OH)SH \underset{}{\overset{-H_2O, H_2S}{\longrightarrow}}$$

$$R-CH(SH)_2 \underset{}{\overset{-H_2S}{\rightleftharpoons}} R-CHS \underset{-S}{\overset{H_2S}{\longrightarrow}} R-CH_2SH \qquad (7\text{-}20)$$

The increase in pulp yield may reach 8% on dry wood basis, but requires high pressures (> 1000 kPa) and a large excess of hydrogen sulfide (ca. 10% of wood). Only a fraction of the hydrogen sulfide (1–2% of wood) is consumed and the rest is recoverable. Table 7-7 illustrates the influence of some oxidizing and reducing agents on the carbohydrate yield of kraft pulp. Stabilization with anthraquinone is dealt with in Section 7.3.7.

7.3.7 Sulfur-Free Pulping

In order to be able to reduce the pollution load of the pulp mills, attempts have been made to diminish the use of sulfur chemicals even if their complete elimination is difficult. Sulfur contaminants are easily introduced into the system, e.g., in connection with the use of oil as fuel, and already traces of them give rise to odor.

Alkaline solutions containing oxygen can be used for the removal of lignin from softwoods, but the delignification in this system is quite unselective. Better results have been obtained for hardwoods using pressurized oxygen at low alkalinities, e.g., sodium carbonate and sodium hydrogen carbonate solutions. Interest has also been directed toward a two-stage oxygen pulping system ("soda-oxygen pulping"). Here, wood chips are first subjected to

TABLE 7-7. **Reductive and Oxidative Stabilization of Softwood Carbohydrates during Kraft Pulping**

Method/addition	Yield of polysaccharide (% of wood)			
	Cellulose	Glucomannan	Glucuronoxylan	Total
Normal sulfate pulping	35	4	5	44
Oxidation/polysulfide (4% S)	36	9	5	50
Reduction/NaBH₄	36	12	4	52
Reduction/H₂S	36	9	4	49

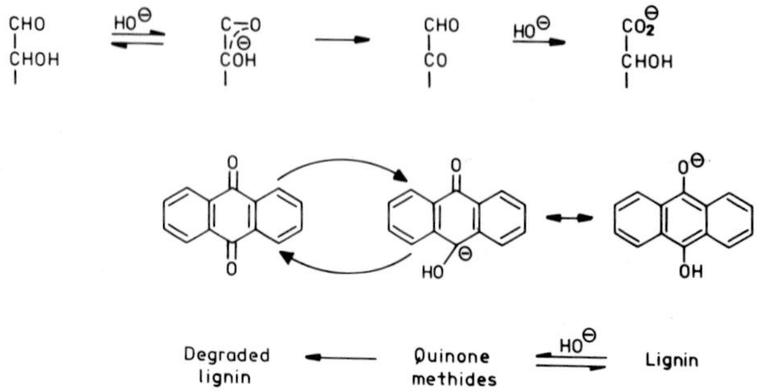

Fig. 7-36. Anthraquinone–anthrahydroquinone reactions with carbohydrates and lignin.

soda pulping and then fiberized mechanically. The final delignification is carried out in the presence of alkali and oxygen.

Sulfide in the kraft cooking liquor can be replaced, at least partly, by anthraquinone (AQ) or similar compounds which possess a marked capability of accelerating the delignification while at the same time stabilizing the polysaccharides toward alkaline degradation according to the mechanism illustrated in Fig. 7-36. At moderate temperatures AQ is reduced to anthrahydroquinone (AHQ) by the polysaccharide end groups, which, in turn, are oxidized to alkali-stable aldonic acid groups. The reduced species or AHQ now acts as an effective cleaving agent with regard to the lignin β-aryl ether linkages in free phenolic phenylpropane units and is simultaneously oxidized to AQ (Fig. 7-37). The partly depolymerized lignin is further degraded by sodium hydroxide at elevated temperature. As a result of this reduction–oxidation cycle, additions as low as 0.01% AQ of the dry wood weight markedly improve the delignification. Depending on the wood species, conditions, desired effect, etc., up to 0.5% may be used.

7.3.8 Resin Reactions

In kraft pulping, the resin and fatty acids which are either free or liberated in the hydrolysis of fats and waxes are dissolved as sodium salts ("soaps") in the cooking liquor. Especially the resin acid salts are effective emulgators

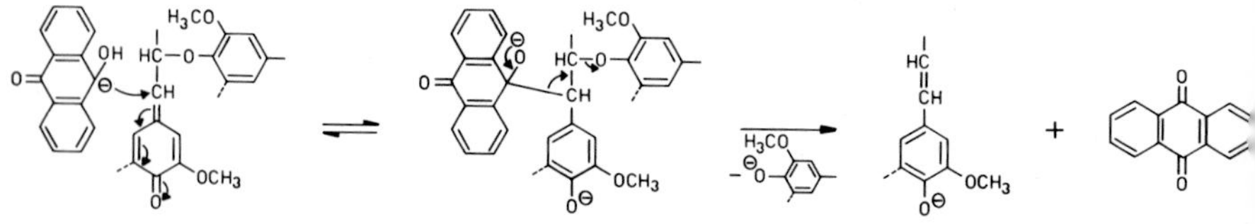

Fig. 7-37. Cleavage of β-aryl ether bonds in alkaline media by anthrahydroquinone with regeneration of anthraquinone.

facilitating the removal of neutral and lipophilic substances. The risk for "pitch problems" is especially great in the case of hardwoods because their resin contains substances with very low solubility and also due to morphological factors. Addition of tall soap to the cook greatly facilitates the removal of resin.

Wood resin is considerably changed during the pulping process. The fatty acid esters of sterols and triterpenoid alcohols in hardwoods (waxes) are saponified very slowly. Unsaturated compounds, e.g., fatty acids, resin acids, and other higher terpenoids, are polymerized to high molecular weight compounds, which also give rise to "pitch problems."

7.3.9 The Composition of Black Liquor

The volatile wood extractives, consisting mainly of lower terpenes, are recovered during kraft pulping from the digester relief condensates (turpentine). The resin acids and fatty acids are afterward recovered as tall soap. After acidification with sulfuric acid the resulting tall oil is finally purified and fractionated by distillation (cf. Section 10.3.1). The remaining kraft spent liquor (black liquor) contains organic constituents in the form of lignin and carbohydrate degradation products (Table 7-8). For their utilization, see Sections 10.3.1 and 10.3.2.

The average molecular weight of the lignin fraction is relatively high (Table 7-9), although minor amounts of low molecular weight degradation products, e.g., guaiacol, vanillin, vanillic acid, and acetoguaiacone, are also present. Most of the lignin (ca. 50%) can be precipitated after neutralization

TABLE 7-8. Typical Composition of the Organic Material in Pine Kraft Spent Liquor

Component	Content (% of dry solids)	Composition (% of hydroxy acids)
Lignin	47	
Hydroxy acids	28	
Lactic		15
2-Hydroxybutanoic		5
2,5-Dihydroxypentanoic		4
Xyloisosaccharinic		5
α-Glucoisosaccharinic		15
β-Glucoisosaccharinic		36
Others		20
Formic acid	7	
Acetic acid	4	
Extractives	5	
Other compounds	9	

TABLE 7-9.   Some Properties of Softwood Björkman Lignin and Kraft Lignin[a]

	Björkman lignin	Kraft lignin
Molecular weight ($\overline{M}_n$)	8000-10000	3000-5000
Polydispersity ($\overline{M}_w/\overline{M}_n$)	2-3	3-4
Sulfur content (%)	—	1-3
Functional groups[b]		
Hydroxyl groups	120	120
Guaiacyl OH	30	60
Catechol OH	—	12
Aliphatic OH	90	48
Carboxyl groups	5	16
Carbonyl groups	20	15
Double bonds		
Coniferyl type	7	—
Stilbene	—	7
Benzyl alcohol or ether		
containing structures		
Noncyclic	42	6
Phenyl coumaran	11	3
Pinoresinol	10	5

[a] See also Marton (1971).
[b] Per 100 C_6C_3 units.

of the black liquor with carbon dioxide (pH 8-9) under simultaneous liberation of the phenolic hydroxyl groups. A further addition of strong acid, e.g., sulfuric acid, results in the liberation of the residual and stronger acidic groups (carboxyls) and more lignin is precipitated (10-30%).

7.3.10   Recovery and Conversation of Kraft Cooking Chemicals

In the kraft process large amounts of comparatively expensive cooking chemicals are used which has necessitated the development of an advanced technology for the recovery of these chemicals in combination with the generation of process energy. The total content of solids in pine kraft black liquor leaving the diffusers of filter washers is 15-20%. It contains most of the degraded and dissolved wood material together with the inorganic chemicals. Most of the base has been consumed for the neutralization of the organic acids; sulfur is still predominantly present as hydrosulfide ions. After partial evaporation, the tall oil skimmings are recovered and treated separately (see Section 10.3.1). The content of solids in the concentrated black liquor coming from the multiple-effect evaporators and entering the Tomlinson-type recovery furnace is 50-70%. The inorganic smelt remaining at the bottom of the furnace after combustion contains mainly sodium carbo-

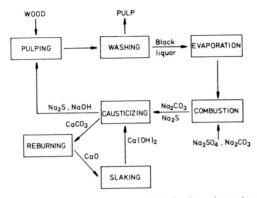

Fig. 7-38. Recovery and conversion of the kraft cooking chemicals.

nate and sodium sulfide. A part of the substance, collected as dust and fumes from the cyclones and electrostatic precipitators at the top of the furnace, consists of sodium sulfate, and is returned to the concentrated black liquor for combustion. The stack gases further contain sodium sulfate in addition to organic sulfur compounds and sulfur dioxide. The losses of chemicals are compensated by adding sodium sulfate to the concentrated black liquor prior to its combustion. To avoid sulfur losses from the stack gases and also to reduce air pollution, the black liquor can be oxidized with air or gaseous oxygen before evaporation and/or combustion. The oxidation converts remaining sulfides to sulfates and mercaptans to disulfides and their further oxidation products.

The smelt leaving the combustion furnace is dissolved in water and the sodium carbonate in the resulting "green liquor" is converted to hydroxide by lime:

$$CaO + H_2O \rightarrow Ca(OH)_2 \tag{7-21}$$

$$Ca(OH)_2 + Na_2CO_3 \rightleftarrows 2\,NaOH + CaCO_3 \tag{7-22}$$

In this procedure, calcium carbonate (lime sludge) is precipitated and separated from the liquid. The remaining solution, consisting mainly of sodium sulfide and sodium hydroxide ("white liquor"), is used as such for cooking. After washing and drying the lime sludge is reburned to give new calcium oxide.

Figure 7-38 illustrates the recovery and conversion of the kraft cooking chemicals. In the case of sulfur-free pulping (soda process, soda-oxygen process, and anthraquinone-alkali pulping) only sodium carbonate is recovered and chemical losses are compensated by adding sodium carbonate. The hydroxide-carbonate and lime cycles are the same as for the kraft process.

References

Enkvist, T., Alfredsson, B., and Martelin, J.-E. (1957). Determinations of the consumption of alkali and sulfur at various stages of sulfate, soda, and alkaline and neutral sulfite digestion of spruce wood. *Sven. Papperstidn.* **60,** 616-620.

Gellerstedt, G. (1976). The reactions of lignin during sulfite pulping. *Sven. Papperstidn.* **79,** 537-543.

Gellerstedt, G., and Gierer, J. (1971). The reactions of lignin during acidic sulphite pulping. *Sven. Papperstidn.* **74,** 117-127.

Ghosh, K. L., Venkatesh, V., Chin, W. J., and Gratzl, J. S. (1977). Quinone additives in soda pulping of hardwoods. *Tappi* **60**(11), 127-131.

Gierer, J. (1970). The reactions of lignin during pulping. *Sven. Papperstidn.* **73,** 571-596.

Gierer, J., Lindeberg, O., and Norén, I. (1979). Alkaline delignification in the presence of anthraquinone/anthrahydroquinone. *Holzforschung* **33,** 213-214.

Goliath, M., and Lindgren, B. O. (1961). Reactions of thiosulphate during sulfite cooking. Part 2. Mechanism of thiosulphate sulphidation of vanillyl alcohol. *Sven. Papperstidn.* **64,** 469-471.

Gustafsson, L., and Teder, A. (1969). Alkalinity in alkaline pulping. *Sven. Papperstidn.* **72,** 795-801.

Hansson, J.-Å. (1970). Sorption of hemicelluloses on cellulose fibres, Part 3. The temperature dependence on sorption of birch xylan and pine glucomannan at kraft pulping conditions. *Sven. Papperstidn.* **73,** 49-53.

Holton, H. H., and Chapman, F. L. (1977). Kraft pulping with anthraquinone. *Tappi* **60**(11), 121-125.

Ingruber, O. V. (1958). The influence of the pH factor in sulphite pulping. *Tappi* **41,** 764-772.

Janson, J., and Sjöström, E. (1964). Behaviour of xylan during sulphite cooking of birchwood. *Sven. Papperstidn.* **67,** 764-771.

Johansson, M. H., and Samuelson, O. (1974). The formation of end groups in cellulose during alkali cooking. *Carbohydr. Res.* **34,** 33-43.

Johansson, M. H., and Samuelson, O. (1977). Alkaline destruction of birch xylan in the light of recent investigations of its structure. *Sven. Papperstidn.* **80,** 519-524.

Kaufmann, Z. (1951). Über die chemischen Vorgänge beim Aufschluss von Holz nach dem Sulfitprozess. Diss., Eidg. Tech. Hochsch. Zürich, Zürich.

Kleppe, P. J. (1970). Kraft pulping. *Tappi* **53,** 35-47.

Landucci, L. L. (1980). Quinones in alkaline pulping. Characterization of an anthrahydroquinone-quinone methide intermediate. *Tappi* **63,** 95-99.

Löwendahl, L., and Samuelson, O. (1977). Carbohydrate stabilization during kraft cooking with addition of anthraquinone. *Sven. Papperstidn.* **80,** 549-551.

Malinen, R., and Sjöström, E. (1975). The formation of carboxylic acids from wood polysaccharides during kraft pulping. *Pap. Puu* **57,** 728-736.

Marton, J. (1971). Reactions in alkaline pulping. *In* "Lignins" (K. V. Sarkanen and C. H. Ludwig, eds.), pp. 639-694. Wiley (Interscience), New York.

Pekkala, O., and Palenius, I. (1973). Hydrogen sulphide pretreatment in alkaline pulping. *Pap. Puu* **55,** 659-668.

Rydholm, S. A. (1965). "Pulping Processes." Wiley (Interscience), New York.

Samuelson, O., and Sjöberg, L.-A. (1972). Oxygen-alkali cooking of wood meal. *Sven. Papperstidn.* **75,** 583-588.

Sanyer, N., and Chidester, G. H. (1963). Manufacture of wood pulp. *In* "The Chemistry of Wood" (B. L. Browning, ed.), pp. 441-534. Wiley (Interscience), New York.

Saukkonen, M., and Palenius, I. (1975). Soda-oxygen pulping of pine wood for different end products. *Tappi* **58**(7), 117-120.

Schöön, N.-H. (1962). Kinetics of the formation of thiosulphate, polythionates and sulphate by the thermal decomposition of sulphite cooking liquors. *Sven. Papperstidn.* **65**, 729-754.

Simonson, R. (1963). The hemicellulose in the sulfate pulping process, Part 1, The isolation of hemicellulose fractions from pine sulfate cooking liquors. *Sven. Papperstidn.* **66**, 839-845.

Simonson, R. (1965). The hemicellulose in the sulfate pulping process, Part 3, The isolation of hemicellulose fractions from birch sulfate cooking liquors. *Sven. Papperstidn.* **68**, 275-280.

Sjöström, E. (1964). Chemical aspects of high-yield pulping processes. *Nor. Skogsind.* **18**, 212-223. (In Swed.)

Sjöström, E. (1977). The behavior of wood polysaccharides during alkaline pulping processes. *Tappi* **60**(9), 151-154.

Sjöström, E., and Enström, B. (1967). Characterization of acidic polysaccharides isolated from different pulps. *Tappi* **50**, 32-36.

Sjöström, E., Haglund, P., and Janson, J. (1962). Changes in cooking liquor composition during sulphite pulping. *Sven. Papperstidn.* **65**, 855-869.

Stone, J. E. (1957). The effective capillary cross-sectional area of wood as a function of pH. *Tappi* **40**. 539-541.

Teder, A. (1969). Some aspects of the chemistry of polysulfide pulping. *Sven. Papperstidn.* **72**, 294-303.

Teder, A., and Tormund, D. (1973). The equilibrium between hydrogen sulfide and sulfide ions in kraft pulping. *Sven. Papperstidn.* **76**, 607-609.

Vroom, K. E. (1957). The "H" factor: A means of expressing cooking times and temperatures as a single variable. *Pulp Pap. Mag. Can.* **58**(3), 228-231.

Wood, J. R., and Goring, D. A. I. (1973). The distribution of lignin in fibres produced by kraft and acid sulphite pulping of spruce wood. *Pulp Pap. Mag. Can.* **74**, T309-T313.

Yllner, S., and Enström, B. (1956). Studies of the adsorption of xylan on cellulose fibres during the sulphate cook, Part 1. *Sven. Papperstidn.* **59**, 229-232.

Chapter 8

PULP BLEACHING

The light absorption (color) of pulp is mainly associated with its lignin component. To reach an acceptable brightness level the residual lignin should thus either be removed from the pulp or, alternatively, freed from strongly light-absorbing groups (chromophores) as completely as practicable. Accordingly, the following two alternatives are possible: (1) *lignin-removing (delignifying)* and (2) *lignin-preserving* bleaching. Delignifying bleaching, which results in both high and reasonably permanent brightness, is applicable to chemical pulps. It is performed in several separate stages, usually with chlorine and chlorine-based chemicals, and also with oxygen.

Because of the dissolution of extractives and such contaminants as rust and bark residues, the cleanliness of the pulp is considerably improved. In the case of dissolving pulps the purpose of bleaching is to remove hemicelluloses besides lignin, while for paper-grade pulps hemicellulose losses should be avoided. Lignin-preserving bleaching, which usually gives only a moderate brightness increase, is the appropriate method for high-yield pulps of mechanical and semichemical types. The most common lignin-preserving bleaching chemicals are sodium dithionite and sodium peroxide.

8.1 Lignin-Removing Bleaching

Although lignin can be removed from the pulp much more selectively by bleaching than by cooking, bleaching chemicals are expensive and the elimination of water-polluting wastes at reasonable costs meets great difficulties. Delignifying by bleaching is thus generally motivated only for the removal of comparatively small residual lignin quantities for which purpose cooking is very unselective and/or ineffective.

Although the consumption of bleaching chemicals is roughly proportional to the content of residual lignin, the relative ease to reach a given brightness level (bleachability) is influenced by a number of factors such as the structure and accessibility of the residual lignin and its distribution across the cell wall. Striking differences exist between the initial brightness of different pulp types: unbleached sulfite pulps can be relatively bright, whereas unbleached kraft pulps have a low brightness and require a more extensive bleaching.

In practice, the bleaching of pulp is accomplished in successive stages, each of them characterized by appropriate chemicals and conditions. The bleaching procedure is chosen with respect to the pulp type in order to attain target brightness with preservation of strength properties (Table 8-1). In recent years increasing attention has been directed toward improvement of the

TABLE 8-1.   Examples on Bleaching Sequences and Conditions[a]

Sequence	Stage	Chemical	Charge[b] (kg/ton pulp)	Consistency (%)	Temp. (°C)	Time (min)
A.	D + C	$ClO_2 + Cl_2$	7 + 63	3.5	30–40	60
	E	NaOH	35	10	60	120
	H	NaOCl + NaOH	12 + 3	8	40	150
	D	ClO_2	15	10–12	70	180
	E	NaOH	8	10–12	65	120
	D	ClO_2	10	10–12	70	240
B.	O[c]	O_2/NaOH	20/20	10–30	100	30
	D + C	$ClO_2 + Cl_2$	3 + 35	3.5	45	30
	E	NaOH	10	10	50	120
	D	ClO_2	10	10	70	180
	E	NaOH	8	10	65	120
	D	ClO_2	5	10	70	180
C.	D + C	$ClO_2 + Cl_2$	8 + 32	3.5	20	45
	E	NaOH	20	10	50	120
	H	NaOCl + NaOH	8 + 2	8	40	180
	D	ClO_2	8	10	75	180

[a] A large variety of conditions and combinations are used, some of them involving sodium peroxide and/or ozone stages (not shown here). Kraft pulps typically require more extensive bleaching (Sequences A and B) than sulfite pulps (Sequence C) (cf. Table 8-2).
[b] NaOCl and ClO_2 as act. Cl.
[c] Inhibitor addition 0.03–1% of pulp (as $MgCO_3$).

entire bleaching technique, and conventional systems are being replaced by new methods. Although the chemistry of bleaching has remained essentially unchanged, the great merit of the new technique of so-called displacement and gas phase bleaching is that the diffusion of chemicals into the pulp phase is considerably accelerated. The bleaching time is thus reduced and the concentration of the organic solids in the resulting bleach liquors is increased. This, in turn, results in more economical processes of handling and elimination of wastes.

8.1.1 General Chemistry of Bleaching

On dissolution of chlorine gas in water, the following equilibria are almost instantly attained:

$$Cl_2 + H_2O \rightleftarrows H^+ + Cl^- + HOCl \qquad (8-1)$$

$$HOCl \rightleftarrows H^+ + ClO^- \qquad (8-2)$$

The equilibrium constants relating to equations (8-1) and (8-2) are:

$$K_1 = [H^+][Cl^-][HOCl] / [Cl_2] \cong 3.9 \times 10^{-4}$$

$$(pK_1 \cong 3.4 \text{ at } 25°C) \qquad (8-3)$$

$$K_2 = [H^+][ClO^-] / [HOCl] \cong 2.9 \times 10^{-8}$$

$$(pK_2 \cong 7.5 \text{ at } 25°C) \qquad (8-4)$$

The K values, which are real constants only at a given concentration since the activity coefficients have been neglected, increase with temperature. An equilibrium diagram based on these values is shown in Fig. 8-1. As can be

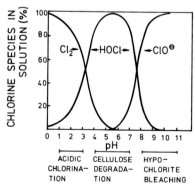

Fig. 8-1. Composition of aqueous chlorine (0.1 M) as a function of pH, based on equilibrium constants corresponding to $K_1 = 3.9 \times 10^{-4}$ and $K_2 = 2.9 \times 10^{-8}$.

seen, undissociated hypochlorous acid is practically the only species present at around pH 5.5. Below this pH the proportion of chlorine is successively increased whereas hypochlorite ions dominate in the alkaline region.

Hypochlorous acid can also be converted into dichlorine monoxide according to the following equation:

$$2\ HOCl \rightleftarrows Cl_2O + H_2O \tag{8-5}$$

and

$$K_3 = [Cl_2O] / [HOCl]^2 < 10^{-2}\ (pK < 2) \tag{8-6}$$

This equilibrium is attained rather slowly and only small amounts of dichlorine monoxide can be present. However, because dichlorine monoxide is extremely reactive, its presence must be considered within the pH region below 7.5.

In the acidic chlorination stage of pulp bleaching, most of the chlorine is consumed rapidly within 5 to 10 minutes. This initial phase is diffusion controlled, and effective mixing of the pulp slurry facilitates the reaction.

Chlorine can react either as a molecular species directly with the organic material or after being decomposed to chlorine radicals. A spontaneous decomposition may take place under the influence of light, when the chlorine radicals formed combine with chloride ions:

$$Cl_2 \rightarrow 2\ Cl\cdot \tag{8-7}$$

$$Cl\cdot + Cl^- \rightarrow Cl_2^-\cdot \tag{8-8}$$

Another mechanism also involving radical formation is the reaction of chlorine with organic compounds:

$$Cl_2 + RH \rightarrow Cl\cdot + R\cdot + HCl \tag{8-9}$$

Although this process is slow, the chlorine radicals once formed will initiate a rapidly accelerating chain reaction:

$$RH + Cl\cdot \rightarrow R\cdot + HCl \tag{8-10}$$

$$R\cdot + Cl_2 \rightarrow RCl + Cl\cdot \tag{8-11}$$

It has been assumed that the reactions of chlorine and hypochlorite with carbohydrates proceed mainly by the radical mechanism because they are retarded in the presence of radical scavengers, such as chlorine dioxide:

$$Cl\cdot + ClO_2\cdot + H_2O \rightarrow Cl^- + ClO_3^- + 2\ H^+ \tag{8-12}$$

The reaction with lignin presumably proceeds via the positive ends of the polarized chlorine and hypochlorous acid molecules ($^{\delta-}Cl-Cl^{\delta+}$ and $^{\delta-}HO-Cl^{\delta+}$).

The chlorination stage of bleaching is usually carried out in the presence of chlorine dioxide, which is either charged first, followed somewhat later by chlorine (D/C bleaching) or, alternatively, chlorine and chlorine dioxide are added simultaneously (D+C bleaching).

The hypochlorite solution is usually prepared by introducing chlorine into an aqueous solution of sodium or calcium hydroxide. The chlorine is disproportionated into chloride and hypochlorite ions according to the equation:

$$2 \text{ NaOH} + \text{Cl}_2 \rightarrow \text{NaOCl} + \text{NaCl} + \text{H}_2\text{O} \qquad (8\text{-}13)$$

On prolonged storage, the hypochlorite ions disproportionate gradually into chloride and chlorate. Hypochlorite solutions may also be decomposed to chloride and oxygen in the presence of heavy metal ions. When hypochlorite is reduced to chloride two oxidation equivalents are consumed. For example, 1 M hypochlorite solution contains $2 \times 35.5 = 71$ g/liter of active chlorine.

Chlorine dioxide, prepared from sodium chlorate by reduction, reacts quite selectively with lignin and is therefore widely used for pulp bleaching. At slightly acidic conditions (pH 4–5) chlorine dioxide can be reduced to chloride ions:

$$\text{ClO}_2 + 4 \text{ H}^+ + 5 \text{ e}^- \rightarrow \text{Cl}^- + 2 \text{ H}_2\text{O} \qquad (8\text{-}14)$$

In this reaction five oxidation equivalents are released. For example, 1 M ClO_2 solution contains $5 \times 35.5 = 177.5$ g/liter of active chlorine. In alkaline media, chlorine dioxide is reduced to chlorite involving a change of only one oxidation equivalent.

The oxidation pathways of chlorine dioxide under actual conditions are complex because a number of species including chlorine, hypochlorous, chlorous, and chloric acids are formed as intermediates. A rapid conversion of chlorine dioxide to chloride and chlorite (chlorous acid, $pK \sim 2.0$) may first take place, followed then by a slow phase during which mainly the chlorite reacts with the pulp components. However, continuous generation of chlorine dioxide during bleaching takes place, for example, by the reaction between chlorite and chlorine (or hypochlorous acid):

$$2 \text{ ClO}_2^- + \text{Cl}_2 \rightarrow 2 \text{ ClO}_2 + 2 \text{ Cl}^- \qquad (8\text{-}15)$$

$$2 \text{ ClO}_2^- + \text{HOCl} \rightarrow 2 \text{ ClO}_2 + \text{Cl}^- + \text{HO}^- \qquad (8\text{-}16)$$

Chlorine and hypochlorous acid may also react with chlorite to form chlorate:

$$\text{ClO}_2^- + \text{Cl}_2 + \text{H}_2\text{O} \rightarrow \text{ClO}_3^- + 2 \text{ Cl}^- + 2 \text{ H}^+ \qquad (8\text{-}17)$$

$$\text{ClO}_2^- + \text{HOCl} \rightarrow \text{ClO}_3^- + \text{Cl}^- + \text{H}^+ \qquad (8\text{-}18)$$

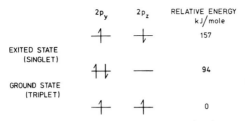

Fig. 8-2. Energy states of oxygen molecule.

Oxygen. Besides chlorine and chlorine containing chemicals, oxygen is used for bleaching. In its ground or triplet state the oxygen molecule has two unpaired electrons in its outer shell, but it can be exited to higher energy levels (singlet state) characterized by two paired or unpaired electrons with an antiparallel spin (Fig. 8-2). Oxygen may be reduced to water in four stages by one-electron transfer that gives rise to peroxy radicals ($HO_2 \cdot$), hydrogen peroxide (H_2O_2), and hydroxyl radicals ($HO \cdot$), or their ionized species as well as their organic counterparts (Fig. 8-3). Because of the presence of all these reactive species the reaction pattern is very complicated and imperfectly understood.

Due to its unpaired electrons oxygen participates in chain reactions (Fig. 8-4). Radicals may be generated by electron abstraction from phenolate ions (in lignin) giving rise to resonance-stabilized phenoxy radicals (Fig. 8-5) or directly by abstraction of hydrogen atoms linked to carbon. Traces of transition metal ions will catalyze the decomposition of the hydroperoxides (or hydrogen peroxide) into extremely reactive alkoxy or hydroxyl radicals which can abstract hydrogen atoms from carbohydrates resulting in their degradation. For delignification, however, reaction sequences involving a two-electron transfer (ionic mechanism), also causing hydroperoxide formation, have been suggested (cf. Fig. 8-13).

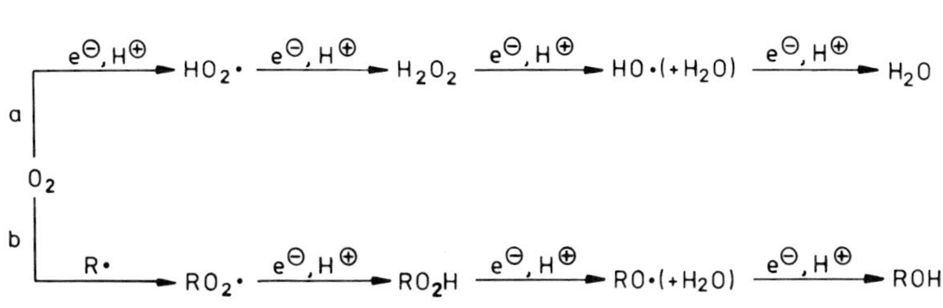

$$
\begin{array}{c}
a \left[\quad \xrightarrow{e^{\ominus}, H^{\oplus}} HO_2 \cdot \xrightarrow{e^{\ominus}, H^{\oplus}} H_2O_2 \xrightarrow{e^{\ominus}, H^{\oplus}} HO \cdot (+H_2O) \xrightarrow{e^{\ominus}, H^{\oplus}} H_2O \right. \\[2mm]
O_2 \\[2mm]
b \left. \quad R \cdot \xrightarrow{\quad} RO_2 \cdot \xrightarrow{e^{\ominus}, H^{\oplus}} RO_2H \xrightarrow{e^{\ominus}, H^{\oplus}} RO \cdot (+H_2O) \xrightarrow{e^{\ominus}, H^{\oplus}} ROH \right.
\end{array}
$$

Fig. 8-3. Reduction of oxygen by a one-electron transfer mechanism. Each of the four reaction steps involves addition of one electron and one proton, giving rise to peroxy radicals, hydrogen peroxide, and hydroxyl radicals as intermediate products (a). The corresponding organic species are formed when oxygen reacts with organic radicals $R \cdot$ (b).

INITIATION

$$RO^{\ominus} + O_2 \longrightarrow RO\cdot + O_2^{\ominus}$$

or

$$RH + O_2 \longrightarrow R\cdot + HO_2^{\cdot}$$

PROPAGATION

$$R\cdot + O_2 \longrightarrow RO_2^{\cdot}$$

$$RO_2^{\cdot} + RH \longrightarrow RO_2H + R\cdot$$

TERMINATION

$$RO\cdot + R\cdot \longrightarrow ROR$$

CATALYSIS

$$RO_2H + Fe^{3\oplus} \longrightarrow RO_2^{\cdot} + Fe^{2\oplus} + H^{\oplus}$$

or

$$RO_2H + Fe^{2\oplus} \longrightarrow RO\cdot + Fe^{3\oplus} + HO^{\ominus}$$

Fig. 8-4. Examples of radical chain reaction initiated by oxygen and decomposition of hydroperoxides catalyzed by transition metal ions. R denotes an organic residue.

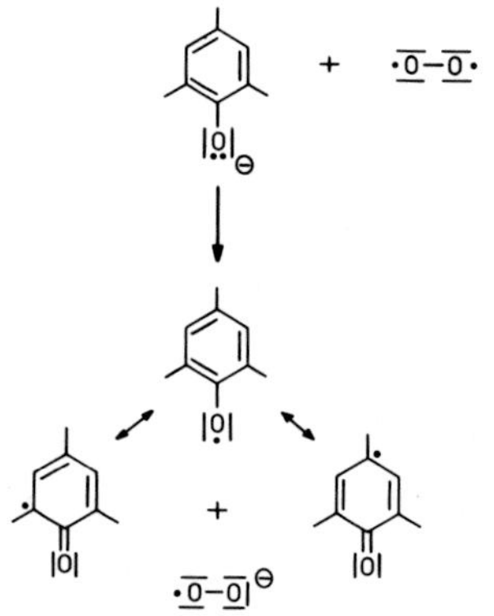

Fig. 8-5. Reaction of oxygen with free phenolic structures leading to the resonance-stabilized phenoxy radicals.

8.1.2   Lignin Reactions

The mechanism responsible for the removal of the residual pulp lignin by bleaching agents is not known in detail and many of the explanations offered so far are only speculative. This can be understood in view of the complex structure of the residual lignin, which also varies depending on the pulping method used. One striking feature is the fact that lignin is removed more easily from sulfite than from kraft pulps (Table 8-2). Although a number of factors, both chemical and morphological, contribute to this difference, it can be noted that sulfite pulps contain sulfonic acid groups and the content of phenolic hydroxyl groups is high in kraft pulps. Because sulfonic acid groups are always dissociated in water, they increase the hydrophilicity of lignin equally over the entire pH range, whereas phenolic hydroxyls contribute to the hydrophilicity especially under strongly alkaline conditions.

Chlorine Stage   Chlorine reacts with lignin primarily by substitution and by oxidation. Small amounts of chlorine are also introduced by addition to the double bonds present in lignin. The electrophilic chlorine reacts with positions carrying a partial negative charge and the reactivity of the aromatic carbon atoms is influenced by the nature of substituent groups. Hydroxyl and methoxyl groups activate the *ortho* and *para* positions toward electrophilic reactions, whereas α-carbonyl and α-carboxyl groups deactivate these positions and are therefore *meta* directors. Depolymerization of lignin may result from two types of reactions, namely, (1) electrophilic side chain displacement and (2) oxidative breakage of aryl ether bonds and decomposition of the aromatic nuclei (Figs. 8-6 and 8-7).

During bleaching about three moles of chlorine are consumed per phenyl

TABLE 8-2.   Example of the Decrease in the Lignin Content after Various Stages during Bleaching of Sulfate and Sulfite Pulps[a]

	Sulfate pulp		Sulfite pulp
Stage	Lignin content[b] (% of pulp)	Stage	Lignin content (% of pulp)
Unbleached	2.4	Unbleached	2.4
C	2.0	C	0.7
E	0.5	E	0.2
C	0.3	D	<0.05
H	<0.05	E	<0.05
D	<0.05	D	<0.05
E	<0.05		

[a] cf. Sjöström and Enström (1966).
[b] Lignin content of unbleached pulp is lower than normally.

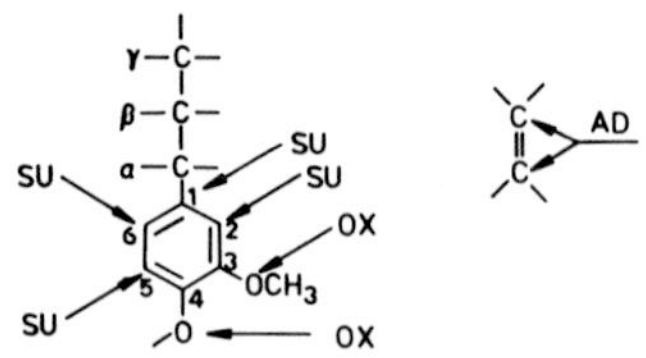

Fig. 8-6.   Reaction sites of chlorine in a guaiacylpropane unit. *Substitution* (SU) which occurs mainly at C-6 and C-5 positions leads to chlorinated lignin products and hydrogen chloride. After substitution at C-1 the side chain is displaced giving rise to lignin fragmentation. *Oxidation* (OX) gives rise to cleavage of methoxyl groups or leads to bond breakage between neighboring units. *Addition* (AD) also leads to the increase in chlorine content, but because of low frequency of double bonds in the side chains it is not a significant reaction.

propane unit. Approximately 50–60% of this chlorine is rapidly consumed in substitution and addition reactions. Oxidation reactions, proceeding rapidly at the initial stage but more slowly later, are responsible for the remaining consumption of chlorine. Lignin is gradually depolymerized and the resulting fragments, rich in carboxyl groups, are dissolved in the bleach liquor.

At a D/C bleaching stage, chlorine dioxide is gradually consumed primarily in oxidation of free phenolic groups, resulting in an improved delignification. If chlorine and chlorine dioxide are added simultaneously (D+C bleaching) the degradation of the carbohydrates is counteracted (see Section 8.1.3), while the delignification is not markedly affected.

Alkaline Extraction Stage.   Extensive lignin dissolution takes place by alkali as a consequence of the conversion of phenolic hydroxyl and carboxyl groups to their more hydrophilic salts. As a result of the ionization of the phenolic groups the light absorption of the liquor from the alkaline extraction stage increases. About 70% of the chlorine substituted in lignin is removed as chloride by alkali with simultaneous liberation of phenolic hydroxyl groups. o-Quinone structures (Fig. 8-7) are possibly subjected to a benzilic acid rearrangement generating furane carboxylic acid structures (Fig. 8-8).

The lignin content of a softwood pulp after the chlorination and extraction stages is approximately 0.5–1.0% (kappa number 4–8). About 3% of the chlorine charged remains in the lignin (chlorine content ca. 10%) after the extraction stage. The brightness of the pulp after prebleaching (C + E stages) is 25–35% (ISO) and the light absorption coefficient (k) at 457 nm is 20–40 m^2/kg. The corresponding k value for the residual lignin is at least 2000

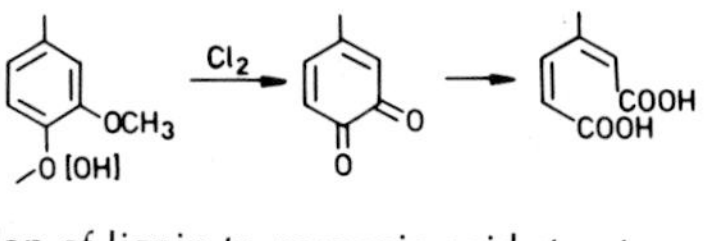

Fig. 8-7.   Chlorine oxidation of lignin to muconic acid structures via o-quinone structures.

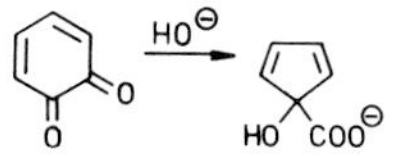

Fig. 8-8. Formation of furan carboxylic acid structures from *o*-quinone structures by a benzilic acid rearrangement (Hardell and Lindgren, 1975).

m²/kg, indicating a strong increase either in the number of chromophoric groups or in their absorptivity. Such groups include conjugated quinoid structures (cf. Fig. 7-30).

Hypochlorite Stage   Because of its negative charge, the hypochlorite ion (ClO⁻) is a nucleophile, in contrast to hypochlorous acid (HOCl) and chlorine (Cl₂) which are electrophiles. Hypochlorite therefore primarily attacks positions carrying a positive charge, especially carbonyl carbon atoms and β-(and δ-)carbons at double bonds conjugated with carbonyl or carboxyl groups (Fig. 8-9). Such groups are formed in lignin during kraft pulping and prebleaching. As a result of these reactions lignin is fragmented to low molecular weight carboxylic acids and dissolved. Because of destruction of chromophoric groups and dissolution of lignin the brightness of the pulp is increased.

Chlorine Dioxide Stage   Prior to the chlorine dioxide stage the pulp has usually undergone several previous bleaching stages, and the chromophore system is very complicated. When the pulp is bleached to high brightness, 90–92% (ISO), the light absorption coefficient at 457 nm decreases to a value of about 0.8–1 m²/kg.

Although the free phenolic structures are oxidized faster, chlorine dioxide also destroys nonphenolic phenyl propane units and double bonds present in the pulp chromophores. After cleavage of the benzene ring various dicarboxylic acids are formed, such as oxalic, muconic, maleic, and fumaric acids in addition to products substituted with chlorine (Fig. 8-10). As a result of depolymerization and formation of carboxyl groups the modified lignin is dissolved during the chlorine dioxide treatment and in the sodium hydroxide extraction stage that usually follows.

Oxygen–Alkali Stage   About one half of the lignin can be removed from a conventional kraft pulp by applying oxygen bleaching as the initial stage. A more complete oxygen delignification by the methods presently used would lead to an extensive degradation of polysaccharides, affecting the pulp quality (see Section 8.1.3). Oxygen and peroxide bleaching have common fea-

Fig. 8-9. Hypochlorite oxidation of a keto group. R and R' denote alkyl or aryl groups. R' can also be Cl.

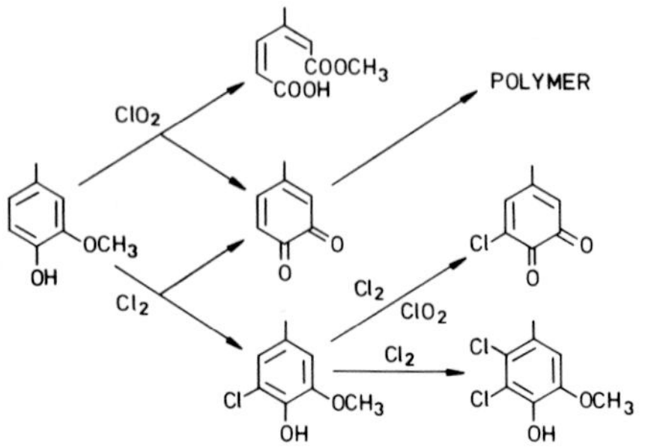

Fig. 8-10. Example of the reactions of a guaiacyl unit with chlorine dioxide and chlorine (Hardell and Lindgren, 1975).

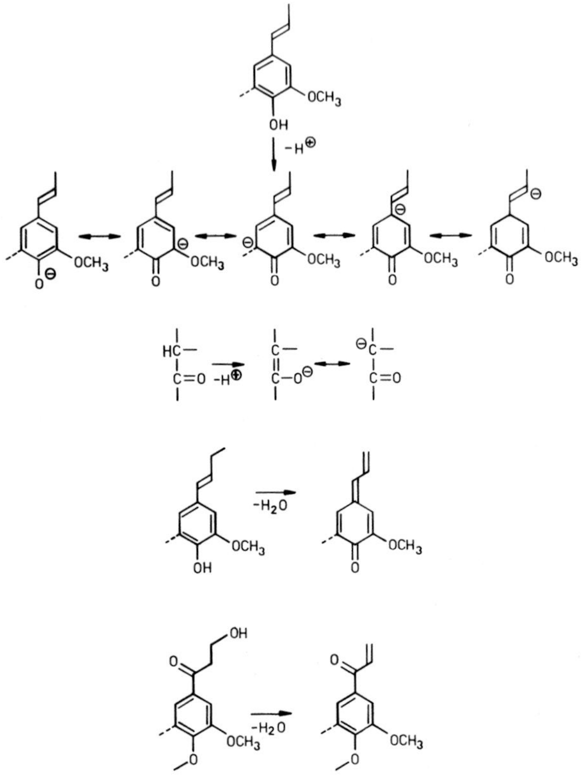

Fig. 8-11. Formation of carbanions and conjugated carbonyl structures (Gierer and Imsgard, 1977).

tures because in both cases the medium is alkaline, and the same reactive species (oxygen and peroxides) are present, although in different proportions. The structures in lignin are converted in alkali to carbanions and conjugated carbonyl structures (Fig. 8-11) giving rise to electron attracting or repelling positions (Fig. 8-12). As an electrophilic reagent, oxygen prefers negatively charged positions, whereas the nucleophilic peroxy anion (HOO⁻) reacts with positively charged positions. Furthermore, the radical sites formed in lignin by hydrogen abstraction with oxygen can also be attacked by another molecule of oxygen.

Both oxygen and hydrogen peroxide react with organic compounds to form hydroperoxides, although via different routes (Fig. 8-13). By action of oxygen, lignin is degraded and chromophoric structures are formed, whereas hydrogen peroxide eliminates chromophores without any marked decomposition and dissolution of the lignin.

8.1.3   Carbohydrate Reactions

Chlorine and Hypochlorite   Although the reaction of chlorine and hypochlorite with lignin is fast, carbohydrates do not remain completely unattacked. Polysaccharides are oxidized to some degree with introduction of carbonyl groups. The polysaccharide chains are subsequently subjected to degradation in the following alkaline extraction stage via β-alkoxy elimination (cf. Fig. 8-20). Glycosidic bonds may also be cleaved directly after

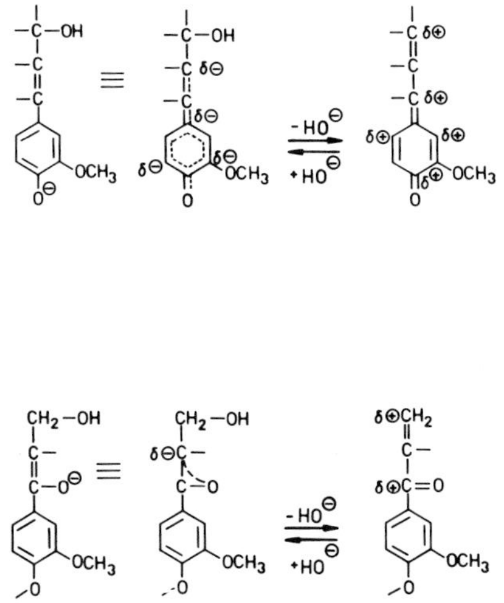

Fig. 8-12.   Sites of electrophilic and nucleophilic attacks by oxygen and hydrogen peroxide, respectively, in phenolic and enolic arylpropane units (Gierer and Imsgard, 1977).

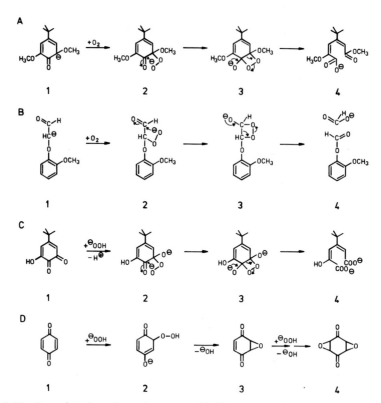

Fig. 8-13. Examples of reactions of oxygen and hydrogen peroxide with lignin model compounds (Gierer and Imsgard, 1977). (A) Carbanion (1) is attacked by electrophilic oxygen, causing formation of hydroperoxide (2) which reacts further via dioxetane (3) to an open-chain structure (4). (B) Structure (4) is obtained from the coupling product (2) of carbanion (1) and oxygen by the C–C bond breakage in dioxetane (3). (C) The reaction of o-quinone (1) with hydrogen peroxide leads to formation of a dicarboxylic acid (4) via hydroperoxide (2) and dioxetane (3) intermediates. (D) The reaction of p-quinone (1) with hydrogen peroxide results in oxirane structures (3) and (4) via a hydroperoxide intermediate (2).

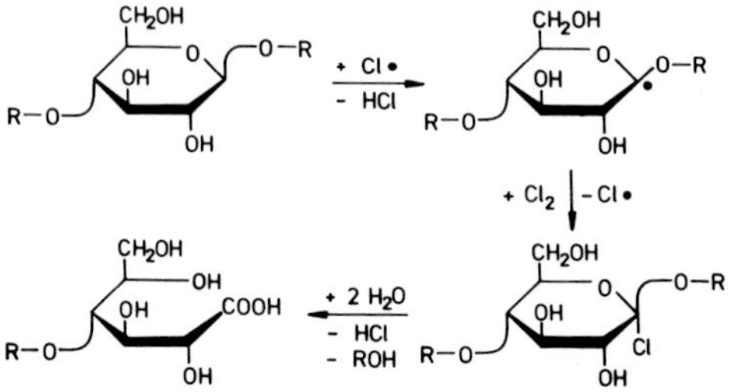

Fig. 8-14. Example of the oxidation of cellulose by chlorine radicals. After oxidation at C-1, the glycosidic bond is cleaved with formation of a gluconic acid end group. R denotes cellulose chain.

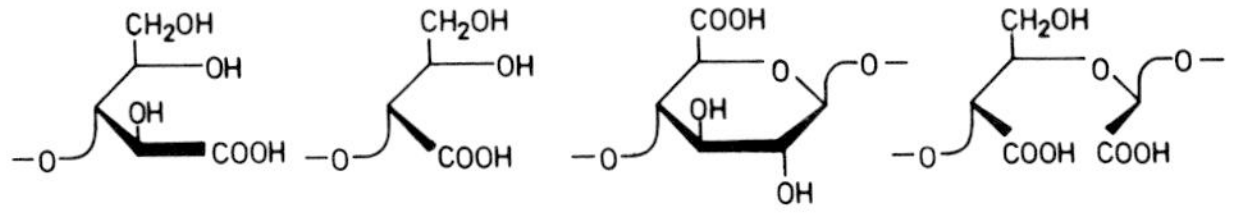

Fig. 8-15. Possible carboxyl group structures formed in cellulose during chlorine and hypochlorite stages. 1, Arabinonic acid; 2, erythronic acid; 3, glucuronic acid; 4, dicarboxylic acid.

oxidation at C-1 by chlorine radicals to yield aldonic acid end groups (Fig. 8-14). In addition, a more extensive degradation yields arabinonic and erythronic acid end groups (Fig. 8-15). Nonterminal carboxyl groups can also be formed without giving rise to any depolymerization. For instance, oxidation at C-6 results in the formation of uronic acid groups.

The oxidation of polysaccharides by hypochlorite proceeds mainly to the carboxylic acid state provided the pH value is above 9. At lower pH values the formation of carbonyl groups dominates, resulting in alkali lability.

Chlorine dioxide is the most selective lignin oxidant among the chlorine-based bleaching chemicals and reacts only very slowly with polysaccharides. Moreover, the stability of the polysaccharides during a chlorination stage is also markedly improved by the presence of chlorine dioxide. Chlorine dioxide then acts as a radical scavenger, inhibiting the radical chain reaction between chlorine and polysaccharides without affecting the chlorination of lignin, which is not a radical reaction.

Oxygen   The viscosity is sharply decreased when a pulp is delignified with oxygen and alkali without any additives (inhibitors). The pulp polysaccharides are attacked by radicals generated by the decomposition of hydroperoxides catalyzed by heavy metal ions or possibly formed directly in the reaction of oxygen with organic material (cf. Fig. 8-4). The peroxide decomposition is partly inhibited by the presence of certain compounds, especially magnesium salts and triethanolamine, capable of deactivating the heavy metals. Other types of inhibitors, acting as radical scavengers or decompos-

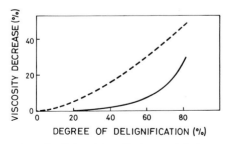

Fig. 8-16. Depolymerization of cellulose (viscosity decrease) during oxygen bleaching of pine kraft pulp. Dotted line, no inhibitor addition; full line, inhibitor added (Mg salt).

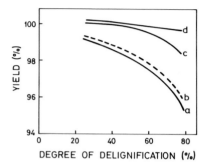

Fig. 8-17. Yield in oxygen bleaching. Full line, inhibitor added; dotted line, no inhibitor addition. (a) Spruce sulfite; (b and c) pine kraft; (d) spruce sulfite pulps (reducing end groups have been eliminated by NaBH₄ reduction).

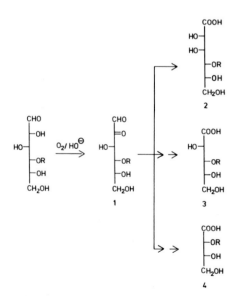

Fig. 8-18. Oxidative stabilization of cellulose through formation of aldonic acid end groups. R is cellulose chain. After oxidative formation of a glucosone intermediate this undergoes a benzilic acid rearrangement strongly favoring formation of mannonic acid end groups (2). In addition, cleavage of C-1 to C-2 and C-2 to C-3 bonds gives rise to the formation of arabinonic acid (3) and erythronic acid (4) end groups. (In this simplified scheme only the main products are shown.)

ing peroxide into inactive species, are also known. Even if various magnesium salts are used in practice, magnesium hydroxide precipitated in the presence of alkali is the active component, primarily because of its ability to strongly adsorb and thereby deactivate the heavy metal ions. As shown in Fig. 8-16, about 50% of the lignin can be removed from a kraft pulp by oxygen–alkali delignification without significant reduction in viscosity, provided magnesium compounds have been added. After further delignification, however, cellulose is gradually depolymerized concomitant with loss in fiber strength.

Oxygen–alkali bleaching is less selective than conventional bleaching, especially when applied to sulfite pulps (Fig. 8-17). The comparatively high carbohydrate losses can be explained by the presence of reducing end groups in sulfite pulp polysaccharides, which initiate the peeling reaction. However, this reaction is to some extent counteracted by oxidation of these end groups to aldonic acid groups (Fig. 8-18). The initial oxidation step involves formation of a glycosone intermediate which, after a benzilic acid rearrangement, is transformed to an epimeric pair of aldonic acids; in the case of cellulose and glucomannan mainly to mannonic acid. Cleavage of carbon–carbon bonds occurs as well, resulting in the formation of arabinonic and erythronic acid end groups predominantly. However, before

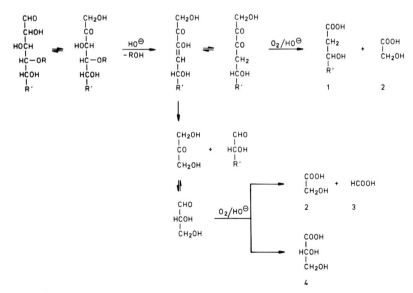

Fig. 8-19. Peeling reaction of polysaccharides during oxygen-alkali treatment. R is polysaccharide chain. Cellulose and glucomannans (R′ = CH$_2$OH): 1, 3,4-Dihydroxybutanoic acid; 2, glycolic acid; 3, formic acid; 4, glyceric acid. Xylan (R′ = H): 1, 3-Hydroxypropanoic acid (2-deoxyglyceric acid); 2, glycolic acid; 3, formic acid; 4, glyceric acid.

this oxidative stabilization has occurred, several sugar units are peeled off from the reducing end of the polysaccharide (Fig. 8-19). Instead of the usual alkaline peeling products (cf. Section 7.3.5) 3,4-dihydroxybutanoic acid and glycolic acid are formed by degradation of the diulose structures by oxygen. Similarly, glyceraldehyde is oxidized to glycolic and glyceric acids.

Since most of the terminal units in kraft pulps are alkali-stable metasaccharinic acid groups in addition to other carboxyl moieties (see Section 7.3.5) the peeling reaction cannot begin here. However, a very harmful reaction of polysaccharides during oxygen bleaching is the oxidation of carbon atoms 2, 3, or 6 to a carbonyl group, which induces a chain cleavage by β-alkoxy elimination. The most common depolymerization reaction occurs after oxidation of the C-2 position (Fig. 8-20). A corresponding oxidation at C-3 leads to the same result because of the migration of the carbonyl group to C-2. The 2-ulose thus formed is easily degraded by β-alkoxy elimination at C-4, resulting in chain cleavage and formation of a new reducing end group. Although chain cleavage at C-1 is also possible (after oxidation at C-3), this does not seem to occur.

Positions C-2 and C-3 can also be oxidized simultaneously. The resulting dicarbonyl derivative is converted to a furanosidic acid unit or directly degraded. Some oxidation of C-6 may also occur, giving rise to chain cleavage at C-4.

8.1.4 Reactions of Extractives

The reactions between bleaching chemicals and resin components result in a variety of degradation and chlorination products, only some of which have been identified. The extractives remaining in the pulp after bleaching also represent a complicated mixture of substances of more or less unknown composition.

The understanding of the reactions of the extractives with bleaching

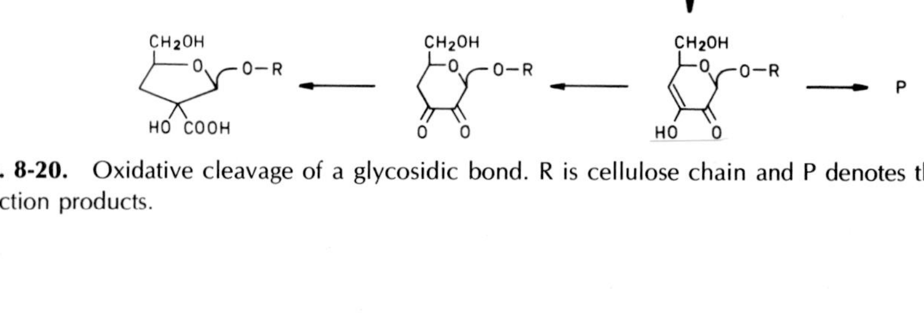

Fig. 8-20. Oxidative cleavage of a glycosidic bond. R is cellulose chain and P denotes the reaction products.

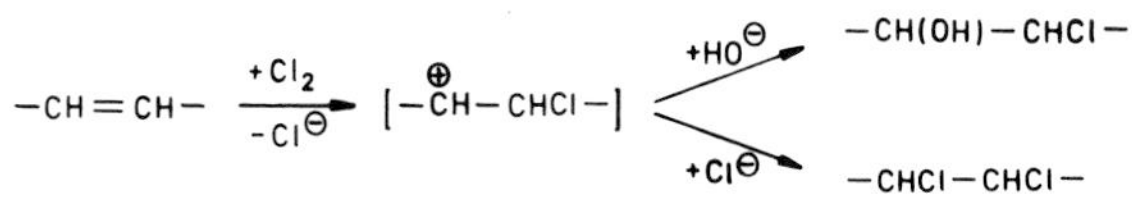

Fig. 8-21. Addition of chlorine to double bonds.

chemicals is mainly based on experiments with some simple resin model compounds and on general knowledge of the reactions of organic compounds with chlorine and chlorine chemicals. *Chlorine* participates primarily in addition reactions with unsaturated compounds, such as fatty acids (Fig. 8-21). Experiments with trioleine (oleic acid triglyceride) show that mainly dichloro compounds are produced after absorption of chlorine into the water-insoluble trioleine phase. The resin left in bleached pulp probably contains largely dichlorinated compounds.

Chlorine dioxide is sometimes used instead of chlorine in the first bleaching stage in order to avoid pitch problems typical of hardwood pulps. Although chlorine dioxide reacts slowly with unsaturated compounds, it probably exerts some oxidizing action on the resin components. Introduction of carboxyl and carbonyl groups also leads to a more effective removal of the resin because of increased water solubility. *Oxygen bleaching* results in a decreased content of extractives and obviously facilitates the deresination in the subsequent stages.

The *last stages of bleaching* (with hypochlorite and chlorine dioxide) have not been found to reduce the resin content significantly.

8.1.5 Spent Bleach Liquors

Disposal of spent liquors from bleach plants represents a serious environmental problem which so far has defied practical solution. While the effluents from the production of unbleached pulp have been reduced drastically by introduction of a closed pulp washing system, a corresponding

TABLE 8-3. Distribution of BOD_7 and Color between the Various Bleaching Stages[a]

	Kraft pulp		Sulfite pulp	
Bleaching stage	BOD_7 (kg O_2/ton)	Color (kg Pt/ton)	BOD_7 (kg O_2/ton)	Color (kg Pt/ton)
C	4.6	7	5.8	25
E	4.8	138	4.7	41
Final bleaching	2.6	9	1.1	0.1
Total	12.0	154	11.6	66

[a] From SSVL (1974).

TABLE 8-4. Composition of the Organic Material in Spent Bleach Liquor from Chlorination and Alkali Extraction Stages (Pine Kraft Pulp)[a]

	Relative composition (%)	
Component	Chlorination	Alkali stage
Chlorinated lignin products	67[b]	75[c]
Carbohydrates	1	5
Volatile acids	3	3
Nonvolatile acids	2	4
Methanol	27	1
Carbon dioxide	—	12
Total amount (kg/ton pulp)		
All compounds	22	47
Compounds with		
M < 1000	15	9

[a] From SSVL (1977) and Pfister and Sjöström (1979).
[b] Chlorine content about 22%.
[c] Chlorine content about 10%.

reduction of the bleach effluents has not been possible. These effluents are therefore responsible for the major part of the total pollution load in a modern pulp mill. Table 8-3 shows the distribution of color and biological oxygen demand (BOD) between the various bleaching stages.

The load from bleach plants has been reduced by the following arrangements: (1) Oxygen bleaching, enabling the effluents from this stage to be combined with cooking spent liquors and thus entering the chemical regeneration system. (2) Countercurrent flow systems, reducing the total volume of effluents and increasing the concentration of organic solids. (3) The closed-cycle bleached kraft pulp mill according to Rapson and Reave.* (4) External processes for the purification of the bleach effluents.

The major part of the dissolved material in bleaching originates from prebleaching (O and CE stages). In the case of kraft pulp, the most extensive dissolution takes place during the extraction stage, whereas even the initial chlorine treatment removes considerable amounts of lignin from sulfite pulp

*In this system the water requirement is minimized by countercurrent washing through the bleaching stages and the resulting effluent is used instead of fresh water for washing of the unbleached pulp. This effluent which contains the solids both from bleaching and cooking is introduced into the kraft chemical recovery system. Because of recycling, the bleaching chemicals are accumulated in the system. To overcome these difficulties chlorine, chlorine dioxide, and sodium hydroxide with given proportions are used for bleaching (mainly chlorine dioxide) to result in the formation of sodium chloride predominantly, which is continuously removed from the white liquor by evaporation and crystallization and reconverted to bleaching chemicals.

(cf. Table 8-2). Considerable amounts of carbohydrates are removed in the hot alkali refinement stage applied to the bleaching of dissolving pulp. The spent bleach liquors are complicated mixtures of reaction products, only some of which have been identified. A simple classification according to the type of compounds present has been made in Table 8-4. As can be seen, lignin degradation products dominate in the spent liquors from chlorination and alkali extraction. Because of extensive demethylation of lignin, considerable amounts of methanol are present in the chlorination liquors. The amount of low molecular chlorinated aromatics is very low, but since the most toxic compounds are included in this fraction, clarification of its complicated composition has been the subject of intensive research.

8.2 Lignin-Preserving Bleaching

A variety of structures giving rise to absorption of visible light are present in chemimechanical pulps, which are the pulp types usually subjected to lignin-preserving bleaching. Such chromophores and potential chromophores (leucochromophores) include phenols and catechols in combination with unsaturated systems of styrene, stilbene, diphenylmethane, and butadiene structures. The standard redox potentials for some quinone structures in the residual lignin are probably of the order $E^0 = 0.7$– 0.9 V and they can thus be reduced relatively easily to the corresponding hydroquinone structures (Fig. 8-22). In principle, chromophores and leucochromophores can be eliminated by either reducing or oxidizing agents. Sulfite, dithionite, and borohydride are used for reduction:

$$H_2SO_3 + H_2O \rightleftharpoons SO_4^{2-} + 4\ H^+ + 2\ e^- \ (E^0 = -0.20\ V) \qquad (8\text{-}19)$$

$$S_2O_4^{2-} + 4\ HO^- \rightleftharpoons 2\ SO_3^{2-} + 2\ H_2O + 2\ e^- \ (E^0 = 1.12\ V) \qquad (8\text{-}20)$$

$$BH_4^- + 8\ HO^- \rightleftharpoons B(OH)_4^- + 4\ H_2O + 8\ e^- \ (E^0 = 1.24\ V) \qquad (8\text{-}21)$$

Although the pulp chromophores are eliminated only partially, dithionite and borohydride are obviously capable of reducing quinones as indicated by the normal potential values. However, hydroquinones (leucochromophores) are easily reoxidized to quinones (chromophores) in the presence of oxygen and light. The quinoid groups in lignin are degraded by alkaline peroxide, leading to a more stable product.

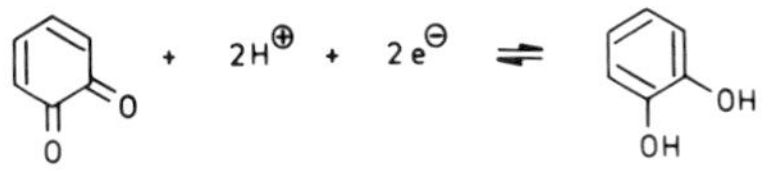

Fig. 8-22. Reduction of o-quinone to catechol.

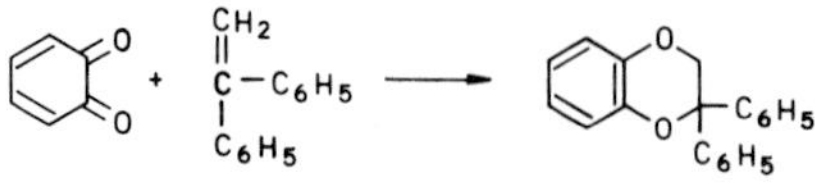

Fig. 8-23. Addition bleaching: an example of the Diels-Alder type reaction (Gierer, 1969).

An interesting possibility for improving the brightness of pulp, although not practicable, is the destruction of conjugated double bond systems by addition reactions of the Diels-Alder type, for example, as shown in Fig. 8-23.

8.2.1 Reducing Bleaching

The use of sodium or zinc dithionite ("hydrosulfite") represents the most common type of reducing bleaching. Usually 0.5–1% sodium dithionite of the dry pulp weight is added and the bleaching is allowed to proceed at 50°–60°C for 1–2 hours. Because ions of heavy metals catalyze the decomposition of dithionite solutions, complexing agents, such as EDTA, are added as stabilizers.

During bleaching, dithionite is decomposed to hydrogen sulfite or sulfite according to equation (8-20). Because of the consumption of hydroxyl ions, the pH decreases. Dithionite may also disproportionate into thiosulfate and hydrogen sulfite:

$$2\ S_2O_4^{2-} + H_2O \rightarrow S_2O_3^{2-} + 2\ HSO_3^{-} \tag{8-22}$$

In the presence of air, dithionite is oxidized:

$$S_2O_4^{2-} + H_2O + O_2 \rightarrow HSO_3^{-} + HSO_4^{-} \tag{8-23}$$

The oxidation of dithionite is accelerated by increasing the hydroxyl ion concentration.

Theoretical considerations reveal that the bleaching effect of dithionite reaches a maximum at pH 8–9 at temperatures of 20°–60°C. However, because oxygen cannot be excluded from the system, a lower pH region is used in practice (pH 5–6) to avoid the oxidation of dithionite. If zinc dithionite is used, the best effect is reached at a still lower pH value.

8.2.2 Oxidative Bleaching

Lignin-preserving oxidative bleaching is almost exclusively carried out with sodium peroxide. Hydrogen peroxide is a weak acid, which is dissociated according to the following equation:

$$H_2O_2 + H_2O \rightleftarrows HO_2^{-} + H_3O^{+} \tag{8-24}$$

$$K = [HO_2^{-}][H_3O^{+}] / [H_2O_2] \cong 10^{-12}$$

$$(pK \cong 12\ at\ 50°–70°C) \tag{8-25}$$

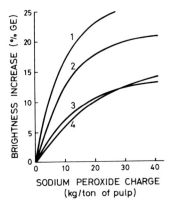

Fig. 8-24. Response of various pulp types to peroxide bleaching. 1, Semichemical (cold alkali) hardwood; 2, acid sulfite softwood; 3, mechanical spruce; 4, semichemical (neutral sulfite) hardwood pulps (Rydholm, 1965).

Alkaline conditions are used to produce peroxide anions (HO_2^-) which are the active bleaching species. It must be remembered that peroxide is readily decomposed in strongly alkaline solutions to oxygen according to the following equation:

$$2\ HO_2^- \rightarrow O_2 + 2\ HO^- \qquad (8\text{-}26)$$

Accordingly, in practice peroxide bleaching is usually performed at an initial pH value of about 11 (final pH $\cong$ 9). The temperature is kept at 50°-60°C for 3-4 hours. Sodium and magnesium silicates are added as stabilizers to prevent decomposition of peroxide, which is catalyzed by heavy metal ions.

The brightening effect attained depends on the peroxide charge, as shown in Fig. 8-24. Peroxide bleaching results in a comparatively large brightness increase, sometimes about 25% (ISO), which is much more than can be achieved with dithionite. However, the response of various pulp types to peroxide bleaching varies greatly.

References

Chang, H.-M., and Allan, G. G. (1971). Oxidation. In "Lignins" (K. V. Sarkanen and C. H. Ludwig, eds.), pp. 433-485. Wiley (Interscience), New York.

Dence, C. W. (1971). Halogenation and nitration. In "Lignins" (K. V. Sarkanen and C. H. Ludwig, eds.), pp. 373-432. Wiley (Interscience), New York.

Gierer, J. (1969). Possibilities of brightness-preserving pulping. *Scand. Symp. Lignin-preserv. Bleach.*, Oslo. (In Swed.), p. 11.

Gierer, J., and Imsgard, F. (1977). The reactions of lignins with oxygen and hydrogen peroxide in alkaline media. *Sven. Pappertidn.* **80,** 510-518.

Hardell, H.-L., and Lindgren, B. O. (1975). Chemical aspects of bleaching kraft pulp with halogen-based chemicals. Part I. *Commun. Swed. For. Prod. Res. Lab. Ser. B* No. 349. (In Swed.)

Lindgren, B. O. (1978). Chemical aspects of bleaching kraft pulps with halogen-based chemicals. Part 2. *Commun. Swed. For. Prod. Res. Lab., Ser. B* No. 504. (In Swed.)

Lindgren, B., and Norin, T. (1969). The chemistry of extractives. *Sven. Pappertidn.* **72,** 143–153. (In Swed.)

Malinen, R. (1975). Behaviour of wood polysaccharides during oxygen–alkali delignification. *Pap. Puu* **57,** 193–204.

Malinen, R., and Sjöström, E. (1972). Studies on the reactions of carbohydrates during oxygen bleaching. Part I. Oxidative alkaline degradation of cellobiose. *Pap. Puu* **54,** 451–468.

Norrström, H. (1972). Light absorption of pulp and pulp components. *Sven. Papperstidn.* **75,** 891–899.

O'Neil, F. W., Sarkanen, K., and Schuber, J. (1962). Bleaching. *In* "Pulp and Paper Science and Technology" (C. E. Libby, ed.), Vol. 1, pp. 346–374. McGraw-Hill, New York.

Pfister, K., and Sjöström, E. (1979). Characterization of spent bleaching liquors. Part 6. Composition of material dissolved during chlorination and alkali extraction (OCE sequence). *Pap. Puu* **61,** 619–622.

Rydholm, S. A. (1965). "Pulping Processes." Wiley (Interscience), New York.

Samuelson, O. (1970). Abbau von Cellulose bei verschiedenen Bleichmethoden. *Papier (Darmstadt)* **24,** 671–678.

Singh, R. P., ed. (1979). "The Bleaching of Pulp," 3rd ed. Tech. Assoc. Pulp Pap. Ind., Atlanta, Georgia.

Sjöström, E. (1980). The chemistry of oxygen delignification. *EUCEPA Symp., Helsinki,* Vol. 1:4.

Sjöström, E., and Enström, B. (1966). Spectophotometric determination of the residual lignin in pulp after dissolution in cadoxen. *Sven. Pappertidn.* **69,** 469–476.

Sjöström, E., and Välttilä, O. (1972, 1978). Inhibition of carbohydrate degradation during oxygen bleaching. Part I. Comparison of various additives. Part II. The catalytic activity of transition metals and the effect of magnesium and triethanolamine. *Pap. Puu* **54,** 695–705; **60,** 37–43.

SSVL Environmental Care Project (1974). "Technical Summary," p. 62. Stockholm. (In Swed.)

SSVL Environmental Care Project No. 7 (1977). "Chloride in Recovery Systems," Final report, p. 23. Stockholm. (In Swed.)

Chapter 9

CELLULOSE DERIVATIVES

9.1 Reactivity and Accessibility of Cellulose

Each β-D-glucopyranose unit within the cellulose chain has three reactive hydroxyl groups, two secondary (HO-2 and HO-3) and one primary (HO-6). Although several factors are involved, one prerequisite for etherification of cellulose is the ionization of the hydroxyl groups. Owing to the inductive effects of neighboring substituents the acidity and the tendency for dissociation is increased according to the series: HO-6 < HO-3 < HO-2. It can therefore be understood why HO-2 is, as a rule, most readily etherified in comparison with the other hydroxyl groups. After substitution of HO-2 the acidity of HO-3 is usually increased which results in its higher reactivity. As concerns esterification, the primary hydroxyl group (HO-6) possesses the highest reactivity.

Another important factor to be considered in the reactions of cellulose concerns the accessibility, which means the relative ease by which the hydroxyl groups can be reached by the reactants. For instance, being least sterically hindered, HO-6 groups show higher reactivity toward bulky substituents than do the other hydroxyl groups.

The morphology of cellulose has a profound effect on its reactivity. The hydroxyl groups located in the amorphous regions are highly accessible and react readily, whereas those in crystalline regions with close packing and strong interchain bonding can be completely inaccessible. The degree of crystallinity depends on the origin of the cellulose preparation (Table 9-1). A preswelling of the cellulose is necessary in both etherifications (alkali) and esterifications (acids).

TABLE 9-1. Degree of Crystallinity of Some Cellulose Samples Measured by Various Techniques[a]

Technique	Cotton	Mercerized cotton	Wood pulps	Regenerated cellulose
Physical				
X-ray diffraction	0.73	0.51	0.60	0.35
Density	0.64	0.36	0.50	0.35
Adsorption and chemical swelling				
Deuteration or moisture regain	0.58	0.41	0.45	0.25
Acid hydrolysis	0.90	0.80	0.83	0.70
Periodate oxidation	0.92	0.90	0.92	0.80
Iodine sorption	0.87	0.68	0.85	0.60
Formylation	0.79	0.65	0.75	0.35
Nonswelling chemical methods				
Chromic acid oxidation	0.997	0.66–0.60[b]	—	—
Thallation	0.996	0.69–0.42[b]	—	—

[a] From Wadsworth and Cuculo (1978).
[b] Mercerization followed by solvent exchange.

The effect of the fine structure on the reactivity of cellulose is demonstrated in Fig. 9-1. When samples of cotton yarn are treated with strong swelling agents to destroy crystallinity, the reactivity toward acetylation is substantially increased especially if drying is omitted. Drastic drying conditions cause extensive interchain hydrogen bonding, thus reducing the accessibility of the hydroxyl groups.

In the strongly swollen or soluble state of cellulose all the hydroxyl groups are accessible to the reactant molecules. However, owing to the random

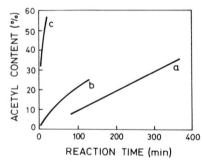

Fig. 9-1. Effect of crystallinity and hydrogen bonding on the acetylation of cotton fibers (Demint and Hoffpauir, 1957). (a) Original fibers. (b) Crystallinity has been destroyed by ethylene amine treatment. Subsequent drying has resulted in the formation of hydrogen bonds. (c) Crystallinity has been destroyed as above but because drying has been omitted no hydrogen bonds have been formed.

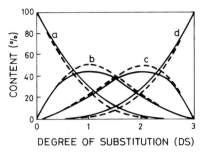

Fig. 9-2. Theoretical distribution of unsubstituted (a), monosubstituted (b), disubstituted (c), and trisubstituted (d) glucose units in cellulose ethers (Timell, 1950). ———, Ratio of rate constants 1:1:1 (HO-2:HO-3:HO-6); ----, ratio of rate constants 5:1:2.

nature of the reaction a homogeneous product is obtained only by complete substitution of the hydroxyl groups (degree of substitution, DS = 3). At any degree of substitution lower than 3 the reaction leads to random sequences of units consisting of the following components: (1) unreacted glucose units, (2) three monosubstituted units (2-; 3-; and 6-derivatives), (3) three disubstituted units (2,6-; 3,6-; and 2,3-derivatives), and (4) the fully substituted units (2,3,6-trisubstituted derivative). The statistical overall distribution of substituents as a function of the degree of substitution (DS) calculated theoretically is shown in Fig. 9-2 (methylation). The curves, which are in good agreement with experimental findings, show that moderate differences in reactivity do not essentially influence the overall distribution of the substituents.

9.2   Swelling and Dissolution of Cellulose

Cellulose swells in different solvents. The extent of swelling depends on the solvent as well as on the nature of the cellulose sample. In the case of native cellulose with fibrous structure more or less drastic morphological changes take place depending on whether the swelling is *interfibrillar* or *intrafibrillar*. More generally the differences caused are distinguished in terms of *intercrystalline* and *intracrystalline* swelling. In the former case the swelling agent enters only into the disordered (amorphous) regions of the cellulose microfibrils and between them, whereas in the latter case the ordered (crystalline) regions are penetrated.

When bone-dry cellulose fibers are exposed to humidity, they adsorb water and the cross section of the fibers is increased because of swelling. At a 100% relative humidity this swelling corresponds roughly to a 25% increase in the fiber diameter. An additional 25% increase in swelling takes place when the fibers are immersed in water. In the longitudinal direction the dimensional change is very small.

The water retention of cellulose fibers at a given relative humidity varies depending on whether the equilibration has taken place by desorption or adsorption (hysteresis). The water uptake also continuously decreases after repeated drying and moistening of the fibers. Additional factors influencing the ability of pulp fibers to swell are their chemical composition, such as their hemicellulose and lignin content.

Cellulose swells in electrolyte solutions because of the penetration of hydrated ions which require more space than the water molecules.

Intracrystalline swelling can be accomplished by concentrated solutions of strong bases or acids and also of some salts. This type of swelling can be either *limited* or *unlimited*. In the former case the swelling agent combines with the ordered cellulose in certain stoichiometric proportions but does not completely destroy the interfibrillar bonding. The latter type refers to cases where the swelling agent is bulky and forms complexes with cellulose thus resulting in breakage of the adjacent bonds and separation of the chains so that gradual dissolution occurs.

The ability of inorganic salt solutions to swell and even dissolve cellulose is usually related to the lyotropic series for the solvated ions but the mechanism is complicated.

Alkali is not capable of dissolving native cellulose. Only depolymerized cellulose fragments with a low degree of polymerization are alkali soluble. Certain quaternary ammonium compounds are more effective resulting in full solubility. A mixture of dimethyl sulfoxide and paraformaldehyde (DMSO-PF) has interesting properties as a cellulose solvent. However, its effect depends at least partly on the formation of a hydroxymethylcellulose derivative. The most important cellulose solvents are metal complexes of

TABLE 9-2. Molecular Formulas and Properties of Common Cellulose Solvents

Abbreviation	Formula	Properties
Schweizer's solution (Cuoxam)	$[Cu(NH_3)_4](OH)_2$	Dark blue. Extensive depolymerization of cellulose in the presence of oxygen.
CED (Cuen)	$[Cu(en)_2](OH)_2$[a]	Dark blue. Depolymerization of cellulose in the presence of oxygen.
Cadoxen	$[Cd(en)_3](OH)_2$	Colorless, useful for optical measurements. Cellulose shows good stability in this solvent.
EWNN	$[FeT_3]Na_6$[b]	Greenish. Cellulose shows good stability in this solvent.

[a] en is ethylenediamine.
[b] T is tartrate.

organic bases; common are cupriethylenediamine (CED) and cadmium ethylenediamine (Cadoxen) (see Table 9-2). Even if other mechanisms are operating, their dissolving ability entails formation of a complex with the two secondary hydroxyl groups in cellulose and with breaking of hydrogen bonds. These solvents are used in connection with the studies of cellulose polymer properties, such as viscosity measurements. Because the solutions are alkaline, cellulose can be depolymerized in the presence of oxygen.

9.3 Swelling Complexes—Alkali Celluloses

When the penetrating agent causes intracrystalline swelling, the X-ray diagram of the cellulose is changed indicating the formation of a cellulose-swelling agent complex. This complex is formed only at a given concentration of the swelling agent. Although extensive swelling can be achieved in solutions of various acids and salts, evidence of definite complexes is often lacking.

The most important swelling complexes of cellulose are those with sodium hydroxide although corresponding addition compounds are formed also with other inorganic and organic bases. The alkali celluloses are compounds with given stoichiometric relations between alkali and cellulose. They are hence usually classified as addition compounds even if their reactions refer to hydrates of alkoxides. For instance, alkali cellulose reacts with alkyl halides in analogy with alkoxides during the Williamson type of ether formation (cf. Section 9.6).

Alkali celluloses are extremely important intermediates because they exhibit a markedly enhanced reactivity compared with original cellulose. The reagents can penetrate more easily into the swollen cellulose structure and thus react with the hydroxyl groups. For instance, preparation of alkali cellulose, named mercerization after its inventor John Mercer (1844), is an important step when producing cellulose xanthate, from which viscose fibers and cellophane are prepared.

When cellulose fibers are mercerized in 12–18% sodium hydroxide solution, the original cellulose (cellulose I) is transformed into cellulose II and the unit cell dimensions are changed. This transformation, taking place at somewhat different concentrations depending on the origin of the sample, can be followed by X-ray measurements (Fig. 9-3).

So-called inclusion compounds can also be prepared from cellulose. According to this technique water-swollen cellulose is dried via a series of solvents and the final solvent used is trapped inside the structure on drying. Although the embedded organic solvent has no stoichiometric relation to the cellulose, it has a property of activating cellulose for organic reactions. Such

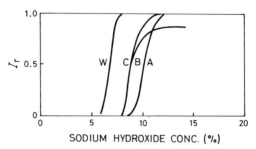

SODIUM HYDROXIDE CONC. (%)

Fig. 9-3. Transition of cellulose I (I_r = 0) to cellulose II (I_r = 1) during cold alkali treatment of wood cellulose (W), cotton cellulose (C), bacterial cellulose (B), and animal cellulose (A) (Rånby, 1952).

inclusion compounds can be prepared from water-swollen cellulose after successive replacement of a series of solvents with decreasing polarity and solubility. As entrapping and activating organic solvents both highly volatile liquids (hexane, benzene, toluene, etc.) and less volatile compounds (e.g., fatty acids) can be used.

9.4   Esters of Inorganic Acids

Cellulose is esterified with certain inorganic acids such as nitric acid, sulfuric acid, and phosphoric acid. A prerequisite is that the acids used can bring about a strong swelling thus penetrating throughout the cellulose structure. The esterification can be considered as a typical equilibrium reaction in which an alcohol and acid react to form ester and water. Of the inorganic esters cellulose nitrate is the only important commercial product.

9.4.1   Cellulose Nitrate

Cellulose nitrate is usually prepared in nitrating acid mixtures containing besides nitric acid sulfuric acid as a catalyst. The first reaction step involves generation of the nitronium ion ($NO_2^{\oplus}$):

$$HONO_2 + 2\,H_2SO_4 \rightleftarrows NO_2^{\oplus} + H_3O^{\oplus} + 2\,HSO_4^{\ominus} \qquad (9\text{-}1)$$

This reaction is an acid–base equilibrium in which sulfuric acid is the acid and the weaker nitric acid is the base, so that this kind of dissociation can take place instead of the formation of usual $H^{\oplus}$ and $NO_3^{\ominus}$ ions. In the next step the electrophilic nitronium ion attacks the hydroxyl groups of the cellulose:

$$NO_2^{\oplus} + HO-Cell \underset{\longleftarrow}{\overset{\longrightarrow}{\quad}} NO_2-\overset{\oplus}{OH}-Cell \underset{\longleftarrow}{\overset{\longrightarrow}{\quad}} NO_2-O-Cell + H^{\oplus} \qquad (9\text{-}2)$$

TABLE 9-3.   Commercial Grades of Cellulose Nitrate[a]

DS	Solvents	Applications
1.9–2.0	Ethanol	Plastics
1.9–2.3	Esters, ethanol, ether–alcohol	Lacquers
2.0–2.3	Esters	Films, cements
2.4–2.8	Acetone	Explosives

[a] From Rånby and Rydholm (1956).

Esterification is retarded by the formation of water which must be removed from the system to force the reaction to completion.

The nitric acid concentration in the nitration acid mixture is usually 20–25%. The degree of nitration can be regulated by changes in the water content. Examples of the solubility and use of cellulose nitrates are given in Table 9-3. As a by-product in the nitration process some cellulose sulfate is also formed (see below). Because this results in instability of the cellulose nitrate, the sulfate groups must be removed by various treatments and the sulfuric acid formed removed by washing.

9.4.2   Cellulose Sulfate

Cellulose sulfates can be prepared by using a variety of reagent combinations (Table 9-4). The active agent is sulfur trioxide (SO_3), present in fuming sulfuric acid or generated according to the following acid–base equilibrium between sulfuric acid molecules:

$$2\,H_2SO_4 \rightleftarrows H_3O^{\oplus} + HSO_4^{\ominus} + SO_3 \tag{9-3}$$

The strongly electrophilic sulfur trioxide is then added to the hydroxyl group and the intermediate oxonium ion is decomposed to sulfate ester and proton:

$$Cell\!-\!OH + SO_3 \rightleftarrows \left[\begin{array}{c} H \\ Cell\!-\!O\!-\!SO_3^{\ominus} \\ \oplus \end{array} \right] \rightleftarrows Cell\!-\!O\!-\!SO_3^{\ominus} + H^{\oplus} \tag{9-4}$$

An alternative mechanism is:

$$Cell\!-\!OH + H_2SO_4 \rightleftarrows \left[\begin{array}{c} H \\ Cell\!-\!OH \\ \oplus \end{array} \right] + HSO_4^{\ominus} \rightarrow Cell\!-\!O\!-\!SO_3^{\ominus} + H_3O^{\oplus} \tag{9-5}$$

Because only one valence of sulfur is occupied for ester formation, the product is acid. Cellulose sulfates are water soluble and can be used as thickening agents.

TABLE 9-4.   Suitable Reagents for Preparing Cellulose Sulfate

Sulfuric acid/ethanol, propanol, butanol
Fuming sulfuric acid/sulfur trioxide
Sulfur trioxide/sulfur dioxide, dimethylformamide, carbon disulfide
Chlorosulfonic acid/sulfur dioxide, pyridine

9.4.3   Other Inorganic Cellulose Esters

Considerable interest has been directed to the preparation of cellulose phosphates because of their flame retarding properties and potential use in textiles. Phosphorylation can be accomplished in several ways, e.g., by heating cellulose at high temperatures with molten urea and phosphoric acid. Other phosphor-containing esters of cellulose include phosphites, phosphinates, and phosphonites. In addition, boric acid esters have been prepared.

9.5   Esters of Organic Acids

9.5.1   Cellulose Acetate

Cellulose acetate has replaced cellulose nitrate in many products, for example, in safety-type photographic films. When a solution of cellulose acetate in acetone is passed through the fine holes of a spinneret and the solvent evaporates, solid filaments are produced. *Acetate rayon* is prepared from threads of these filaments. Some applications and solvents of commercial cellulose acetate grades are summarized in Table 9-5.

Because acetylation of cellulose proceeds in a heterogeneous system, the reaction rate is controlled by the diffusion of the reagents into the fiber structure. The quality of the cellulose raw material used for acetate rayon is

TABLE 9-5.   Commercial Grades of Cellulose Acetate[a]

DS	Solvents	Applications
1.8–1.9	Water–propanol–chloroform	Composite fabrics
2.2–2.3	Acetone	Lacquers, plastics
2.3–2.4	Acetone	Acetate rayon
2.5–2.6	Acetone	X-ray and safety films
2.8–2.9	Methylene chloride–ethanol	Insulating foils
2.9–3.0	Methylene chloride	Fabrics

[a] From Rånby and Rydholm (1956).

of great importance. Although cotton linters fulfill high quality requirements, most cellulose acetate today is produced from wood pulps because of their favorable price and constant availability. Both sulfite and prehydrolyzed kraft pulps are used. Some quality requirements are shown in Table 9-6.

Cellulose acetate is usually produced by the so-called solution process with exception of the fully acetylated end product (triacetate). In the solution process the pulp is first pretreated with acetic acid in the presence of a catalyst, usually sulfuric acid. The purpose of this activation step is to swell the fibers and increase their reactivity as well as to decrease the DP to a suitable level. Acetylation is then performed after addition of acetic anhydride and catalytic amounts of sulfuric acid in the presence of acetic acid. After full acetylation the final triacetate obtained is dissolved. This "primary" acetate is usually partially deacetylated in aqueous acetic acid solution to obtain a "secondary" acetate with a lower DS of about 2 to 2.5.

The fibrous acetylation process is performed in the presence of a suitable liquid, such as benzene, in which the reaction product is insoluble and which thereby retains the fiber form. For fibrous acetylation vapor-phase treatment with acetic anhydride can also be used. Besides sulfuric acid, perchloric acid and zinc chloride have been used as catalysts.

The acid-catalyzed acetylation of cellulose proceeds according to the following reaction formula:

$$
\begin{array}{ccc}
\underset{\substack{\text{O} \\ \|}}{CH_3-C-OCOCH_3} \;\xrightarrow[\;\; +H^{\oplus}\;\;]{} &
\left[\begin{array}{c}
\overset{\oplus}{C}\overset{OH}{\underset{\|}{}} \\
CH_3-C-OCOCH_3 \\
\updownarrow \\
\overset{OH}{\underset{|}{}} \\
CH_3-\underset{\oplus}{C}-OCOCH_3
\end{array}\right] \;
\xrightarrow[+\,Cell-OH]{} &
\begin{array}{c}
CH_3-\overset{OH}{\underset{\underset{\oplus}{HO-Cell}}{C\mathord{-}OCOCH_3}} \\
\Big\updownarrow \;\; {-CH_3COOH \atop -H^{\oplus}} \\
\underset{\substack{\| \\ CH_3-C-O-Cell}}{O}
\end{array}
\end{array}
\tag{9-6}
$$

TABLE 9-6.   Typical Specifications for Acetylation Grade Pulp[a]

α-Cellulose (%)	>95.6
Pentosans (%)	<2.1
Intrinsic viscosity (dm³/kg)	550-750
Ether extractable (%)	<0.15
Ash (%)	<0.08
Iron (mg/kg)	<10

[a] From Malm (1961).

After protonation of the acetic anhydride the electrophilic carbonium ion formed is added to a nucleophilic hydroxyl oxygen atom of the cellulose. This intermediate is then decomposed into cellulose acetate and acetic acid with liberation of a proton.

9.5.2   Other Esters of Organic Acids

A number of various esters of cellulose are known, for example, propionate and butyrate and mixed esters such as acetate–butyrate, propionate–isobutyrate, and propionate–valerate. The mixed esters have found use in plastic composites when good grease- and water-repelling properties are required.

Cellulose can also be esterified by aromatic acids. However, derivatives of any importance are only the cellulose cinnamic and salicylic acid esters. A number of nitrogen-containing esters are also known, for example, cellulose dialkyl diaminoacetate, cellulose acetate-N,N-dimethylaminoacetate, and cellulose propionate-3-morpholine butyrate. Because of the presence of basic substituents these derivatives, although water insoluble, can be dissolved in acidic solutions. Such derivatives have found use as surface coatings in photographic films and in tablets for pharmaceutical purposes.

9.6   Ethers

Cellulose ethers can be prepared by treating alkali cellulose with a number of various reagents including alkyl or aryl halides (or sulfates), alkene oxides, and unsaturated compounds activated by electron-attracting groups. A variety of products of considerable commercial importance has been developed for different uses (Table 9-7). Most of the cellulose ethers are water soluble and they generally possess similar properties, but because

TABLE 9-7.   Types of Commercial Cellulose Ethers

Cellulose ether	Reagent	Solvent	DS
Methylcellulose	Methyl chloride, dimethyl sulfate	Water	1.5–2.4
Ethylcellulose	Ethyl chloride	Organic solvents	2.3–2.6
Carboxymethylcellulose	Sodium chloroacetate	Water	0.5–1.2
Hydroxyethylcellulose	Ethylene oxide	Water	1.3–3.0[a]
Cyanoethylcellulose	Acrylonitrile	Organic solvents	2.0

[a] Molar substitution.

of specific characteristics they complete rather than compete with each other. The water solubility is generally attained at very low degrees of substitution. With hydrophobic substituents and a high DS cellulose ethers become soluble in organic solvents.

9.6.1 Alkyl Ethers

The simplest representatives of cellulose ethers are the corresponding alkyl derivatives. The most common representatives manufactured industrially are methyl- and ethylcellulose. Methylcellulose is soluble in cold water when the DS is 1.4 to 2.0, whereas nearly completely substituted products (DS 2.4–2.8) are insoluble in water but soluble in organic solvents.

Alkali cellulose from cotton linters or wood pulp, usually prepared in a way similar to the first step in the viscose process (see Section 9.7), is used as raw material. Alkylation is carried out by using alkyl chlorides. The reaction proceeds according to the S_N2 mechanism (bimolecular nucleophilic substitution):

$$\text{Cell—OH} + \text{HO}^\ominus \rightleftarrows \text{Cell—O}^\ominus + H_2O \qquad (9\text{-}7)$$

$$\text{Cell—O}^\ominus + \text{R—Cl} \rightleftarrows \text{Cell—OR} + \text{Cl}^\ominus$$
$$(R = CH_3 \text{ or } C_2H_5) \qquad (9\text{-}8)$$

As a by-product methanol or ethanol is formed:

$$\text{RCl} + \text{HO}^\ominus \rightleftarrows \text{ROH} + \text{Cl}^\ominus \qquad (9\text{-}9)$$

which then reacts with alkyl chloride to form dimethyl or diethyl ether:

$$\text{ROH} + \text{HO}^\ominus \rightleftarrows \text{RO}^\ominus + H_2O \qquad (9\text{-}10)$$

$$\text{RO}^\ominus + \text{R—Cl} \rightleftarrows \text{ROR} + \text{Cl}^\ominus \qquad (9\text{-}11)$$

Methylcellulose solutions generally form gels at higher temperatures. The gelation temperature is increased when hydroxyethyl or hydroxypropyl groups are introduced into the methylcellulose (cf. Section 9.6.2). Hydroxyethylmethylcellulose and hydroxypropylmethylcellulose are prepared industrially by the reaction of alkali cellulose first with ethylene oxide or propylene oxide and then with methyl chloride. Similarly, hydroxyethylethylcellulose is prepared by consecutive ethylene oxide and ethyl chloride treatments. Cellulose ethers with both methyl and ethyl groups have also been manufactured.

At a viscosity range exceeding 600 mN·sec·m^{-2} (cP) methylcellulose solutions are pseudoplastic which means that the apparent viscosity decreases with increasing shear rate. Solutions of low viscosities again tend to be thixotropic, resulting in a decreased viscosity with increasing shear times.

Alkyl ethers of cellulose are used as additives for a variety of products. These applications include agricultural products (thickening and dispersing seeds and powders), food products (stabilizer and thickening agents), ceramics (agents improving viscosity and shrink resistance), technochemical products (additives improving wet-rub resistance and flow, etc.), pharmaceutical preparations (tablets, suspensions, emulsions, etc.), cements (control of the setting time), textiles (sizing and coating), paper products, and plywood.

9.6.2 Hydroxyalkyl Ethers

Hydroxyalkyl celluloses are obtained in the reaction of cellulose with alkene oxides or their corresponding chlorohydrins. The reaction is a base-catalyzed S_N2-type substitution, and the reaction rate is proportional to the product [epoxide][Cell—$O^\ominus$]. The commercial preparations include hydroxyethyl- and hydroxypropylcellulose for which ethylene oxide and propylene oxide are used as reagents. Hydroxyethylcellulose is formed according to the following equation:

$$\text{Cell—O}^\ominus + \underset{O}{\overset{H_2C \text{——} CH_2}{\triangle}} \longrightarrow \text{Cell—O—CH}_2\text{CH}_2\text{O}^\ominus \qquad (9\text{-}12)$$

Ethylene oxide can also react with hydroxide ions resulting in the formation of ethylene glycol:

$$\text{HO}^\ominus + \underset{O}{\overset{H_2C \text{——} CH_2}{\triangle}} \longrightarrow \text{CH}_2\text{OH—CH}_2\text{O}^\ominus \qquad (9\text{-}13)$$

In addition, ethylene oxide is polymerized to polyethylene oxide. The terminal primary hydroxyl group of the substituent reacts with additional epoxide to form pendant oxyethylene chains:

$$\text{Cell—O—CH}_2\text{CH}_2\text{OH} + n \underset{O}{\overset{H_2C \text{——} CH_2}{\triangle}} \xrightarrow{\text{HO}^\ominus} \text{Cell—(O—CH}_2\text{CH}_2)_{n+1}\text{OH} \qquad (9\text{-}14)$$

Because of this reaction the degree of substitution (DS) of hydroxyalkylcelluloses is lower than the molar substitution (MS). The ratio MS/DS is a measure of the relative length of side chains. Usually only half of the ethylene oxide reacts with cellulose, the other half is consumed for side reactions.

Water soluble hydroxyethylcelluloses have molar substitutions ranging from 1.5 to 2.5. Hydroxyethylcellulose is used as a thickener for latex paints, for emulsion polymerization of polyvinylacetate, for paper sizing, and for improving the wet strength of paper (together with glyoxal), in ceramic industry, etc.

Hydroxypropylcellulose is applied for similar purposes as hydroxyethylcellulose although its use is more limited. Because hydroxypropyl substitution improves thermoplasticity and solubility in organic solvents, hydroxypropylcellulose can also be used as a thickener for organic solutions. Hydroxypropylation of cellulose does not result in water and alkali soluble products until the MS is close to 4.

9.6.3   Sodium Carboxymethylcellulose

Sodium carboxymethylcellulose (CMC) is the most widely used water-soluble derivative of cellulose. It is prepared from alkali cellulose with sodium chloroacetate as reagent:

$$\text{Cell}-O^{\ominus} + \overset{\frown}{\text{CH}_2\text{COO}^{\ominus}} \longrightarrow \text{Cell}-O-\text{CH}_2\text{COO}^{\ominus} + \text{Cl}^{\ominus} \qquad (9\text{-}15)$$

CMC is manufactured in the DS range from 0.4 to about 1.4. The DP varies from 200 to 1000. The water solubility of CMC is increased when the DS increases. At DS values of 0.6–0.8 good water solubility is attained, whereas preparations with a DS of 0.05–0.25 are soluble only in alkali. Because of the carboxyl groups CMC is a polyelectrolyte. Its pK_a value varies from about 4 to 5, depending on the DS.

CMC can be used in a variety of products such as detergents, foods (as protective colloid and for purposes where high water-binding capacity is required, stabilizers, etc.), ice cream, paper coatings, emulsion paints, drilling fluids, ceramics, pharmaceuticals, and cosmetics.

9.6.4   Cyanoethylcellulose

Cellulose reacts with α,β-unsaturated compounds containing strongly electron-attracting groups to form substituted ethyl ethers. The most common derivative of this type is cyanoethylcellulose. Cyanoethylation requires strongly basic catalysts and is usually carried out in the presence of sodium hydroxide with acrylonitrile as reagent:

$$\text{Cell}-O^{\ominus} + \overset{\delta+}{\text{CH}_2}=\overset{}{\text{CH}}-\text{C}\equiv\overset{\delta-}{\text{N}} \qquad \text{Cell}-\text{OCH}_2\text{CH}_2\text{CN} + \text{HO}^{\ominus}$$

$$\Big\updownarrow \qquad\qquad\qquad \Big\updownarrow +H_2O \qquad (9\text{-}16)$$

$$\text{Cell}-O-\text{CH}_2-\overset{\ominus}{\text{CH}}-\text{C}\equiv\text{N} \longleftarrow \text{Cell}-O-\text{CH}_2-\text{CH}=\text{C}=\text{N}^{\ominus}$$

The cellulose anion formed attacks the positive carbon atom in acrylonitrile to form a resonance-stabilized intermediate anion, which then adds a proton from water to form the product under simultaneous liberation of a hydroxyl ion. All the reaction steps are reversible and because of the regeneration of hydroxyl ions no alkali is consumed. However, the process is accompanied with the consumption of acrylonitrile in several side reactions, such as formation of 3,3'-oxydipropionitrile:

$$2 \ CH_2 = CHCN + H_2O \rightarrow O(CH_2CH_2CN)_2 \qquad (9\text{-}17)$$

Processes have been developed to minimize these side reactions.

Highly cyanoethylated celluloses with a DS of about 2.5 are soluble in polar organic solvents. Because of an unusually high dielectric constant and low dissipation factor they can be used as the resin matrix for the phosphorous and electroluminescent lamps. Cyanoethylated kraft wood pulp with a DS of only about 0.2 is used for making insulation paper for transformers. Cyanoethylated paper has also a good thermal and dimensional stability.

9.7 Cellulose Xanthate

The preparation of viscose *rayon* fibers and *cellophane* proceeds via the xanthate, which therefore is an extremely important derivative of cellulose. Treatment of alkali cellulose with carbon disulfide results in the formation of cellulose xanthate (dithiocarbonate):

$$\text{Cell}-\text{O}^{\ominus} + \overset{S}{\underset{S}{\overset{\|}{C}}} \ \rightleftharpoons \ \text{Cell}-\text{O}-\overset{}{\underset{\underset{S}{\|}}{C}}-S^{\ominus} \qquad (9\text{-}18)$$

In the first step cellulose is treated with 18% sodium hydroxide at 15°–30°C. After removing excess sodium hydroxide from the fibers by pressing the alkali cellulose is shredded and subjected to alkaline ripening to bring the DP down to 200–400. Xanthation is then carried out at 25°–30°C for ca. 3 hours resulting in a DS of approximately 0.5. The cellulose xanthate is dissolved in aqueous sodium hydroxide resulting in an orange-colored viscous liquid, called *viscose*. After ripening the viscose solution is filtered and forced through a spinnerette into an acid (sulfuric acid and salts) bath where the cellulose is regenerated in the form of fine filaments resulting in rayon fibers:

$$\text{Cell}-\text{O}-\underset{\underset{S}{\|}}{C}-S^{\ominus} \overset{H^{\oplus}}{\rightarrow} \text{Cell}-\text{OH} + CS_2 \qquad (9\text{-}19)$$

Cellophane is prepared by pressing the viscose through a narrow slit into an acid bath to form thin sheets.

9.8   Cross-Linking of Cellulose

In order to improve the properties of cellulose fibers mainly for applications in textiles, the cross-linking of cellulose has been studied extensively in heterogeneous systems to maintain the fibrous structure. Usually agents capable of forming cross-links through ether bonds with the hydroxyl groups of cellulose are used; ester cross-links are also possible but not so useful because of their low stability against alkali. New reactive groups can also be introduced into cellulose as coupling points for cross-linking agents.

The most common type of cellulose cross-linking is accomplished by using aldehydes and dialdehydes, such as formaldehyde and glyoxal as reactants. Already at the beginning of this century it was observed that treatment of cellulose with formaldehyde in the presence of acid results in improved wet strength of the fibers. As shown later formaldehyde evidently condenses with the hydroxyl groups of cellulose producing intermolecular ether cross-links, although it is not known to what extent the adjacent secondary hydroxyl groups of a glucopyranose unit participates in this reaction:

$$2\ Cell\!-\!OH + HCHO \xrightarrow{H^{\oplus}} Cell\!-\!O\!-\!CH_2\!-\!O\!-\!Cell \qquad (9\text{-}20)$$

In the 1950s formaldehyde was commercially applied to improve the dimensional stability of rayon fabrics, but it has largely been replaced by other cross-linking agents. The most important technology for cross-linking of cellulose is based on the use of formaldehyde precondensates with amides including ureas, triazines, and carbamates. For example, urea and its derivatives condense with formaldehyde according to the following equation:

$$2\ CH_2O + \underset{\substack{| \\ R}}{HN}\!-\!\underset{\substack{|| \\ O}}{C}\!-\!\underset{\substack{| \\ R}}{NH} \longrightarrow \underset{\substack{| \\ R}}{HOCH_2N}\!-\!\underset{\substack{|| \\ O}}{C}\!-\!\underset{\substack{| \\ R}}{NCH_2OH} \qquad (9\text{-}21)$$

(R = H or substituent)

After impregnation of the cellulose substrate with an aqueous solution containing urea-formaldehyde precondensates and subsequent short heating at 130°-160°C, cross-linking takes place:

$$2\ Cell\!-\!OH + \underset{\substack{| \\ R}}{HOCH_2N}\!-\!\underset{\substack{|| \\ O}}{C}\!-\!\underset{\substack{| \\ R}}{NCH_2OH} \longrightarrow Cell\!-\!\underset{\substack{| \\ R}}{OCH_2N}\!-\!\underset{\substack{|| \\ O}}{C}\!-\!\underset{\substack{| \\ R}}{NCH_2O}\!-\!Cell \qquad (9\text{-}22)$$

Such a treatment of fabrics ("finishing") results in improved dimensional stability, crease resistance, wash-and-wear performance, and durable-press properties, all important for textiles. A variety of agents including cyclic urea derivatives has been applied to find the best effects. Another approach includes the use of activated vinyl compounds and their derivatives such as acrylamides and vinyl ketones. Furthermore, polyfunctional compounds containing oxirane and aziridinyl groups have been used for creation of cross-links into cellulose.

9.9 Grafting on Cellulose

In addition to being the basis of traditional ether and ester derivatives, cellulose can be modified by cross-linking or by preparing so-called graft copolymers. Cross-linking improves certain properties, such as crease and shrink resistance, but because of the three-dimensional network formed the structure of cellulose becomes more rigid. The product is then less suitable for textile purposes because it is more brittle. By grafting, however, polymer branches can be created to the cellulose backbone without destroying the desirable properties of the original cellulose fibers. Graft copolymerization has been applied to a number of cellulosic materials, including cotton, rayon, paper, cellophane, and wood. However, only minor commercial success has so far been attained despite all research.

Although cellulose can be grafted homogeneously using soluble cellulose derivatives or suitable solvents, grafting is usually performed in a heterogeneous system and is greatly influenced by the physical structure of the cellulose. The vast majority of grafting methods involve polymerization of vinyl monomers of type $CH_2{=}CH{-}X$ where X is an inorganic moiety, such as halide, $-CN$, $-NO_2$, or an organic substituent.

The grafting methods can in principle be divided into three categories, namely, radical polymerization, ionic polymerization, and condensation or addition polymerization. Only the first case is discussed in the following since the most common grafting methods belong to this category.

A free radical-initiated grafting reaction can be generated when a vinyl monomer is polymerized in the presence of cellulose. Grafting is initiated by abstraction of a hydrogen atom from cellulose either by a growing chain radical or directly by a radical created from the catalyst. The first pathway is termed *chain transfer*. In both cases an unpaired electron is left on the cellulose chain which initiates grafting. In common with other radical reactions, grafting reactions involve stages of initiation, propagation, and termination as illustrated by the following equations:

Creation of initiator radicals $\quad 1 \rightarrow 2\,R\cdot$ $\hfill$ (9-23)

Initiation of homopolymer radicals* $\quad R\cdot + M \rightarrow RM\cdot$ $\hfill$ (9-24)

Homopolymer chain growth (propagation)· $\quad RM\cdot + M \rightarrow RM_2\cdot$, etc. $\hfill$ (9-25)

Chain transfer to cellulose $\quad RM_{\dot{x}} + Cell\!-\!H \rightarrow RM_xH + Cell\cdot$ $\hfill$ (9-26)

Graft copolymer chain initiation $\quad Cell\cdot + M \rightarrow Cell\!-\!M\cdot$ $\hfill$ (9-27)

Graft side chain growth (propagation) $\quad Cell\!-\!M\cdot + M \rightarrow Cell\!-\!M_2\cdot$, etc. $\hfill$ (9-28)

Homopolymer termination $\quad M_x\cdot + M_{\dot{x}+n} \rightarrow M_{2x+n}$ $\hfill$ (9-29)

Graft side chain termination $\quad Cell\!-\!M_x\cdot + M_{\dot{x}+n} \rightarrow Cell\!-\!M_{2x+n}$ $\hfill$ (9-30)

Cross-linking $\quad Cell\!-\!M_x\cdot + Cell\!-\!M_{\dot{x}+n} \rightarrow Cell\!-\!M_{2x+n}\,Cell$ $\hfill$ (9-31)

For initiation, use is made of peroxides or azo compounds, which can be decomposed to free radicals ($R\cdot$) generating growing monomer radicals (equations 9-23, 9-24, and 9-25). Because of chain transfer to cellulose (equation 9-26) a branch is initiated (equation 9-27) and the side-chain begins to grow (equation 9-28).

Certain compounds, such as those bearing thiol groups (—SH), facilitate chain transfer. Thiol groups, leading to higher grafting yields, can be introduced into the cellulose by reaction with ethylene sulfide:

$$Cell\!-\!OH \;+\; \underset{\displaystyle S}{H_2C\!\!-\!\!-\!\!CH_2} \longrightarrow Cell\!-\!O\!-\!CH_2CH_2SH \qquad (9\text{-}32)$$

After abstraction of hydrogen from the thiol group by a growing chain radical, the chain is terminated under simultaneous generation of a thiol radical:

$$Cell\!-\!O\!-\!CH_2CH_2SH + R\!\leadsto\!\cdot \rightarrow R\!\leadsto\!H + Cell\!-\!O\!-\!CH_2CH_2S\cdot \qquad (9\text{-}33)$$

Formation of graft polymer is then started:

$$Cell\!-\!O\!-\!CH_2CH_2S\cdot + n(CH_2\!=\!CHX) \rightarrow$$

$$Cell\!-\!O\!-\!CH_2CH_2S\!-\!(CH_2\!-\!CHX)_n\cdot \qquad (9\text{-}34)$$

Another method to improve the yield of grafting involves generation of the

*M is a monomer.

initiating species in the swollen cellulose substrate itself. Because the monomer concentration in this system is low, the reacting species have a greater chance to initiate graft copolymerization (equation 9-35) than homopolymerization of the monomers (equation 9-24).

$$\text{Cell—H} + \text{R} \cdot \rightarrow \text{Cell} \cdot + \text{RH} \tag{9-35}$$

To produce radicals the cellulose substrate can be saturated with potassium persulfate. Another method includes impregnation of cellulose with ferrous salts followed by treatment with a monomer solution in the presence of hydrogen peroxide. Hydroxyl radicals are produced in this system according to the following reaction:

$$\text{Fe}^{2\oplus} + \text{H}_2\text{O}_2 \rightarrow \text{Fe}^{3\oplus} + \text{HO}^\ominus + \text{HO} \cdot \tag{9-36}$$

The hydroxyl radicals can react with cellulose, initiating graft copolymerization, or react with monomer, resulting in homopolymerization. A similar redox system is based on the use of ceric ions, which produce radicals by direct oxidation of the cellulose chains and thus initiate graft polymerization:

$$\text{Cell—H} + \text{Ce}^{4\oplus} \rightarrow \text{Cell} \cdot + \text{Ce}^{3\oplus} + \text{H}^\oplus \tag{9-37}$$

$$\text{Cell} \cdot + \text{M} \rightarrow \text{graft copolymer} \tag{9-38}$$

More specific initiators are Mn^{3+} ions in aqueous solution which give efficient grafting of acrylic and methacrylic monomers onto cellulose and very little homopolymer ($\leq 2\%$).

In another method radicalizable groups, such as hydroperoxides, are introduced into cellulose by ozone treatment. The hydroperoxides are decomposed directly or in the presence of reducing agents to radicals which initiate the graft copolymerization:

$$\text{Cell—O}_2\text{H} + \text{Fe}^{2\oplus} \rightarrow \text{Cell—O} \cdot + \text{Fe}^{3\oplus} + \text{HO}^\ominus \tag{9-39}$$

$$\text{Cell—O}_2\text{H} + \text{Fe}^{2\oplus} \rightarrow \text{Cell—O}^\ominus + \text{Fe}^{3\oplus} + \text{HO} \cdot \tag{9-40}$$

$$\text{Cell—O} \cdot + \text{M} \rightarrow \text{graft copolymer} \tag{9-41}$$

$$\text{HO} \cdot + \text{M} \rightarrow \text{homopolymer} \tag{9-42}$$

Both ultraviolet light and high-energy radiation such as gamma rays from a ^{60}Co source or highly accelerated electrons have been used for initiation of grafting. The radical sites created on the cellulose initiate the copolymerization in the presence of vinyl monomers. However, cellulose radicalized by irradiation is degraded because of the cleavage of glucosidic bonds.

Radicals can finally be created by mechanical means, such as milling or by using electrical discharges, but these methods have found only very limited use.

9.10 Cellulose Ion Exchangers and Enzyme Derivatives

Cellulose is suitable for certain applications for which a solid and inert matrix with a large surface area is required. Introduction of acidic or basic substituents into the backbone of cellulose results in cation and anion exchangers, respectively (Table 9-8). Although the cellulose ion exchangers are chemically less stable than the synthetic ion exchange resins and their capacity is relatively low, they are useful particularly for biochemical separation problems involving large molecules such as proteins. Cellulose ion exchangers are commercially available as powder as well as in the form of fiber and paper and they can thus be used for various types of separation techniques including column, paper, and thin-layer chromatography.

Cation Exchange Celluloses. The cation exchange celluloses are of two types: a weakly acidic type containing carboxylic acid groups and a strongly acidic type containing sulfonic acid or phosphoric acid groups. The most common type of a carboxylic acid ion exchanger is carboxymethylcellulose with a low DS giving a cation exchange capacity of 0.4–0.7 meq/g. Products of higher capacity cannot be applied to chromatography unless they are cross-linked to prevent extensive swelling. Other carboxylic acid cellulose cation exchangers can be obtained by oxidation, for example, with nitrogen dioxide which converts the primary hydroxyl groups into carboxyls rather

TABLE 9-8. Cellulose Ion Exchangers

Name	Functional group
Cation exchangers	
Cellulose phosphate (P-cellulose)	$-OPO_3^{2-}$
Sulfoethylcellulose (SE-cellulose)	$-OC_2H_5SO_3^-$
Carboxymethylcellulose (CM-cellulose)	$-OCH_2COO^-$
Oxidized cellulose	$-COO^-$
Anion exchangers	
Aminoethylcellulose (AE-cellulose)	$-OC_2H_4\overset{+}{N}H_3$
Diethylaminoethylcellulose (DEAE-cellulose)	$-OC_2H_4\overset{+}{N}H(C_2H_5)_2$
Triethylaminoethylcellulose (TEAE-cellulose)	$-OC_2H_4\overset{+}{N}(C_2H_5)_3{}^a$
ECTEOLA-cellulose (condensation product of epichlorohydrin, triethanolamine, and cellulose)	

^a Only a part of the functional groups are of this type.

selectively. Sulfoethylcellulose represents a cellulose cation exchanger of strongly acidic type and is prepared by reacting α-chloroethanesulfonic acid with cellulose in the presence of sodium hydroxide:

$$\text{Cell—OH} + \text{ClCH}_2\text{CH}_2\text{SO}_3\text{Na} + \text{NaOH} \rightarrow \text{Cell—OCH}_2\text{CH}_2\text{SO}_3\text{Na} + \text{NaCl} + \text{H}_2\text{O} \quad (9\text{-}43)$$

The capacity of the product is about 0.5 meq/g. Sulfomethylcellulose has been prepared by using monochloromethanesulfonate, made from dichloromethane and sodium sulfite. Phosphonomethylcellulose, a bifunctional cation exchanger, is prepared by reacting disodium chloromethylphosphonate and cellulose:

$$\text{Cell—OH} + \text{ClCH}_2\text{PO}_3\text{Na}_2 + \text{NaOH} \rightarrow \text{Cell—OCH}_2\text{PO}_3\text{Na}_2 + \text{NaCl} + \text{H}_2\text{O} \quad (9\text{-}44)$$

Anion Exchange Celluloses   Aminoethylcellulose, a weakly basic anion exchanger, is prepared in the reaction of cellulose with α-aminoethylsulfuric acid in the presence of sodium hydroxide:

$$\text{Cell—OH} + \text{H}_2\text{NCH}_2\text{CH}_2\text{OSO}_3\text{H} + \text{NaOH} \rightarrow$$
$$\text{Cell—OCH}_2\text{CH}_2\text{NH}_2 + \text{NaHSO}_4 + \text{H}_2\text{O} \quad (9\text{-}45)$$

The capacity is about 0.2 meq/g but can be increased after cross-linking up to about 0.7 meq/g. Diethylaminoethylcellulose (DEAE-cellulose) is made from cotton linters, previously cross-linked with formaldehyde or 1,3-dichloropropanol, by using 2-chlorotriethylamine as reagent. Quarternary cellulose anion exchangers of strongly basic type have also been prepared by reacting DEAE-cellulose with alkyl halides under anhydrous conditions to yield, for example, triethylaminoethylcellulose (TEAE-cellulose). After reaction of cellulose with epichlorohydrin and triethanolamine in the presence of excess alkali, another anion exchanger or so-called ECTEOLA-cellulose can be produced having an ion exchange capacity of 0.3–0.4 meq/g.

In addition to the types mentioned above, a variety of special cellulose ion exchangers have been prepared including materials containing complexing groups showing strong preference for heavy metal ions. Furthermore, specific immunological adsorbents for isolation of antibodies from blood serum have been made by attaching the corresponding antigens to ion exchange celluloses or to unmodified celluloses.

References

Bikales, N. M. (1971). Ethers from α,β-unsaturated compounds. *In* "Cellulose and Cellulose Derivatives" (N. M. Bikales and L. Segal, eds.), Part V, pp. 811–833. Wiley (Interscience), New York.

Bikales, N. M., and Segal, L., eds. (1971). "Cellulose and Cellulose Derivatives," Part V. Wiley (Interscience), New York.

Buytenhuys, F. A., and Bonn, R. (1977). Distribution of substituents in CMC. *Papier (Darmstadt)* **31,** 525-527.

Cassidy, H. G., and Kun, K. A. (1965). "Oxidation-reduction Polymers," pp. 41-52. Wiley (Interscience), New York.

Demint, R. J., and Hoffpauir, C. L. (1957). Influence of pretreatment on the reactivity of cotton as measured by acetylation. *Text. Res. J.* **27,** 290-294.

Gal'braikh, L. S., and Rogovin, Z. A. (1971). Derivatives with unusual functional groups. *In* "Cellulose and Cellulose Derivatives" (N. M. Bikales and L. Segal, eds.), Part V, pp. 877-905. Wiley (Interscience), New York.

Goldman, R., Goldstein, L., and Katchalski, E. (1971). Water-insoluble enzyme derivatives and artificial enzyme membranes. *In* "Biochemical Aspects of Reactions on Solid Supports" (G. R. Stark, ed.), pp. 1-72. Academic Press, New York.

Haines, A. H. (1976). Relative reactivities of hydroxyl groups in carbohydrates. *Adv. Carbohydr. Chem. Biochem.* **33,** 11-109.

Hiatt, G. D., and Rebel, W. J. (1971). Esters. *In* "Cellulose and Cellulose Derivatives" (N. M. Bikales and L. Segal, eds.), Part V, pp. 741-784. Wiley (Interscience), New York.

Malm, C. J. (1961). Pulp for acetylation. *Sven. Papperstidn.* **64,** 740-743.

Mark, H. F., Gaylord, N. G., and Bikales, N. M., eds. (1965). "Encyclopedia of Polymer Science and Technology," Vol. 3, pp. 131-549. Wiley (Interscience), New York.

Mutton, D. B. (1964). Cellulose chemistry. *Pulp Pap. Mag. Can.* **65,** T41-T51.

Rånby, B. (1952). Fine structure and reactions of native cellulose. Ph.D. Thesis, Univ. of Uppsala, Uppsala.

Rånby, B., and Rydholm, S. (1956). Cellulose and cellulose derivatives. *In* "Polymer Processes" (C. Schildknecht, ed.), pp. 351-428. Wiley (Interscience), New York.

Rowland, S. P. (1978). Hydroxyl reactivity and availability in cellulose. *In* "Modified Cellulosics" (R. M. Rowell and R. A. Young, eds.), pp. 147-167. Academic Press, New York.

Rydholm, S. A. (1965). "Pulping Processes," pp. 100-156. Wiley, New York.

Savage, A. B. (1971). Ethers. *In* "Cellulose and Cellulose Derivatives" (N. M. Bikales and L. Segal, eds.), Part V, pp. 785-809. Wiley (Interscience), New York.

Segal, L. (1971). Effect of morphology on reactivity. *In* "Cellulose and Cellulose Derivatives" (N. M. Bikales and L. Segal, eds.), Part V, pp. 719-739. Wiley (Interscience), New York.

Stannett, V. T., and Hopfenberg, H. B. (1971). Graft copolymers. *In* "Cellulose and Cellulose Derivatives" (N. M. Bikales and L. Segal, eds.), Part V, pp. 907-936. Wiley (Interscience), New York.

Tesoro, G. C., and Willard, J. J. (1971). Crosslinked cellulose. *In* "Cellulose and Cellulose Derivatives" (N. M. Bikales and L. Segal, eds.), Part V, pp. 835-875. Wiley (Interscience), New York.

Timell, T. (1950). Studies on cellulose reactions. Ph.D. Thesis, Univ. of Stockholm, Stockholm.

Tripp, V. W. (1971). Measurements of crystallinity. *In* "Cellulose and Cellulose Derivatives" (N. M. Bikales and L. Segal, eds.), Part IV, pp. 305-323. Wiley (Interscience), New York.

Turbak, A. F., ed. (1975). "Cellulose Technology Research," ACS Symposium Series, No. 10. Am. Chem. Soc., Washington, D.C.

Turbak, A. F., ed. (1977). "Solvent Spun Rayon, Modified Cellulose Fibres and Derivatives," ACS Symposium Series, No. 58. Am. Chem. Soc., Washington, D.C.

Wadsworth, L. C., and Cuculo, J. A. (1978). Determination of accessibility and crystallinity of cellulose. *In* "Modified Cellulosics" (R. M. Rowell and R. A. Young, eds.), pp. 117-146. Academic Press, New York.

Ward, K., Jr. (1973). "Chemical Modification of Papermaking Fibers." Dekker, New York.

Chapter 10

CHEMICALS FROM WOOD AND BY-PRODUCTS AFTER PULPING

On a worldwide basis, wood constitutes an enormous, renewable raw material resource (biomass) for production of energy and chemical products. In connection with wood-processing industries large amounts of both solid residues and dissolved material remain as waste. A rational utilization of this organic waste is of utmost importance not only from the pollution point of view, but also because of the necessity of finding substitutes for products based on petroleum and other fossil raw materials, all expensive and limited in supply.

10.1 Wood Chemicals

10.1.1 Extractives

Some pine species can be induced to exude pathological resin by wounding living trees. This exuded *oleoresin* is collected manually. Tar and pitch isolated from this gum was originally used for the protection and tightening of the hulls of wooden ships and for the preservation of ropes ("Naval stores") and gave rise to the so-called *naval stores industry*. Centered in the southeastern longleaf and slash pine areas of the United States this industry has a long history, dating back to the early decades of the seventeenth century. Another naval stores industry was established in the Les Landes region of southwestern France utilizing maritime pine.

Turpentine and Rosin   The major naval stores products are *turpentine*, primarily composed of volatile terpenes, and *rosin*, mainly a mixture of resin acids. Principally the same products, but in different proportions, can be recovered by tapping resin-rich trees or by steam distillation or solvent extraction of wood residues and especially stumps. Today, however, the most important source for turpentine and rosin is the tall oil recovered after pulping of pine wood (see Section 10.3.1).

Conifer needles also contain volatile products which are mainly composed of mono- and sesquiterpenoids. The volatile turpentine constituents are used in the production of flavors and perfumes and in various technical and pharmaceutical products. The turpentine and rosin production of trees can be increased considerably by injecting suspensions of pitch canker fungus or chemicals such as "paraquat" (1,1'-dimethyl-4,4'-bipyridinium salt) which has been found to be very effective.

Wax and Sterols   Bark, needles, and leaves contain a variety of compounds, but so far their commercial utilization has been limited. The nonpolar bark constituents can be extracted with organic solvents and among these a marketable wax product has been developed which is used for different purposes such as a binder or a substitute for carbon wax. The volume of bark in Douglas fir (*Pseudotsuga menziesii*) is especially high compared with that of other North American conifers and therefore gives a high yield of wax. Douglas fir is the dominant timber species in western United States and Canada, and interest has been directed toward the use of bark wastes also because these represent a considerable environmental problem. The wax fraction contains a variety of substances, predominantly mixed esters derived from higher aliphatic alcohols and acids. Suberin and sterols, particularly β-sitosterol, are also recovered. The extracted bark finds use in the manufacture of gaskets.

The interest in sitosterol is related to its potential use in the synthesis of steroid hormones and related pharmaceutical products, such as cortisone derivatives. The bark of birches contains significant amounts of betulinol which is a triterpenoid alcohol. So far no commercial uses have been found for this bark constituent.

Interesting substances possessing juvenile hormone activity against hemipteran bug have been detected in conifers belonging to *Abies* species (+)-Juvabione [methyl ester of (+)-todomatuic acid], isolated from balsam and Japanese fir wood, may have potential use for insecticidal applications. A related compound have been isolated from Douglas fir wood.

Phenolic Compounds   Another group of wood and bark extractives comprises polar phenolic compounds. Among these are the *lignans*. Although markets for lignans have not yet been established, extensive research has been conducted to extract and recover conidendrin which is present in large quantities in the wood of hemlock and spruce species. One potential use of conidendrin is related to its antioxidant properties. Another example is the extraction of heartwood from western red cedar (*Thuja plicata*) which contains lignans related to plicatic acid. Hot water extraction of cedar wood chips or sawdust and subsequent purification for removal of impurities have given plicatic acid in yields corresponding to about 4% of the dry wood. Plicatic acid can be readily converted to derivatives which can be used as antioxidants for food products, including fats and oils.

Among phenolic extractives those having a flavonoid structure are commercially most important. A variety of monomeric as well as oligo- and polymeric phenolic compounds are widely distributed in woody plants, especially in heartwood, barks, leaves, fruits, and roots. Condensed tannins, based on polymerized flavonoids, such as catechin, and tannin extracts are important commercial products, primarily used in the leather and dye industries. Most of the leather in the United States was earlier treated with domestic tannin extracted from hemlock and oak bark or chestnut wood. This tannin has now been almost completely replaced by other vegetable tannin products, isolated mainly from the South American quebracho tree and Acacia trees in Africa. During the 1950s interest was directed in the United States to the isolation of tannins from coniferous bark. The abundant supplies of Douglas fir bark are a potential source of dihydroquercetin. The recovery of tannins from other conifer barks including redwood (*Sequoia sempervirens*) and hemlock have also been studied. Although lacking great commercial success so far, the polymeric polyphenols of bark or so-called phenolic acids (alkali-extracted) and their sulfonates (sulfite-extracted) have found use for drilling oils, agricultural chemicals (metal complexes), and soil stabilization agents (chromate complexes).

10.1.2 Hydrolysis Products

Wood polysaccharides are mostly hydrolyzed by acids but enzymes can also be used (Fig. 10-1). Acid hydrolysis is carried out with either strong or dilute acids. Process applications have been developed in which aqueous hydrochloric or sulfuric acid solutions are passed through the wood material at elevated temperatures (140°–180°C) with continuous withdrawal of hydrolyzate to prevent decomposition of the liberated monosaccharides. The hydrolyzate containing 4–6% sugars is neutralized and concentrated by evaporation. An average yield of 50 to 55% sugars of the dry wood weight is typical although it varies, depending on the raw material. Considerable

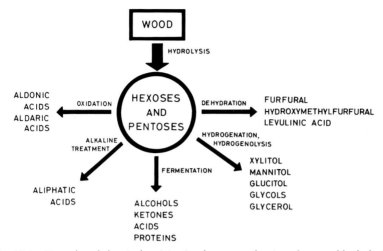

Fig. 10-1. Examples of chemicals originating from sugar fraction after wood hydrolysis.

amounts of cellulose remain unhydrolyzed when dilute acid solutions are used, while the hemicelluloses are converted almost completely to monosaccharides. Interest has also been directed toward enzymic processes for production of glucose from wastes containing cellulose. In this connection new and effective cellulase enzymes from the tropical fungus, *Tricoderma viride,* show promise. One great difficulty, however, is the incomplete hydrolysis of cellulose by enzymes because of the restricted accessibility of its ordered regions.

A considerable wood hydrolysis industry with rather old traditions is located in the Soviet Union. The main fermentation product based on hexoses in wood hydrolyzates is ethyl alcohol, but pentoses and aliphatic acids can also be utilized in the production of proteins (see Section 10.2.3). A variety of chemicals, including ethylene, ethylene oxide, acetaldehyde, and acetic acid, can be produced from ethyl alcohol. One interesting future application of ethyl alcohol concerns its use as a motor fuel mixed with gasoline (gasohol).

The most important pentose is xylose which can be produced from hardwoods by mild acid hydrolysis. Reduction of xylose gives xylitol, an interesting sweetener because of its ability to prevent dental caries. Industrial production of xylitol from birch wood hydrolyzates started in the 1970s in Finland. An interesting process was developed in which an ion exclusion separation technique is applied for purification and separation of xylose and xylitol from other impurities. Among the corresponding reduction products of hexoses, mannitol, which is also a natural product, has found some use. It can be separated from other alditols by crystallization. Under more drastic conditions pentoses and hexoses are converted as a result of hydrogenolysis into glycerol, ethylene glycol, and other glycols.

At higher temperatures, pentoses are further converted by acids to furfural and hexoses under similar conditions to hydroxymethylfurfural, which is further degraded to levulinic acid (cf. Section 2.5.4). Furfural can be produced from xylan-rich raw materials such as oat hulls, corncobs, or hardwood by two-stage hydrolysis via xylose as intermediate product or directly in one stage. The starting material is heated in a digester, usually in the presence of an acid catalyst. The steam introduced is continuously withdrawn from the digester (water-steam distillation). The condensed vapors consist mainly of furfural and water which are separated by azeotropic distillation. Finally, the furfural and water layers are separated and furfural is dried. Furfural is used as an industrial solvent and as a starting material for production of various chemicals and polymers.

Hydroxymethylfurfural is not volatile by steam. It is prepared from hexoses in the presence of an acid catalyst by short heat treatment to avoid further degradation to levulinic acid. After recovery by solvent extraction hydroxymethylfurfural is purified by distillation. Levulinic acid can be prepared in good yield from hexose-based polysaccharides by heating with acids. In this reaction formic acid is liberated and levulinic acid is easily lactonized to form α- and β-angelica lactones (Fig. 2-31).

The solid wood residue remaining after acid hydrolysis is mainly composed of lignin, which is more or less condensed depending on the conditions during the acid treatment. It can be used as a fuel or possibly be subjected to various treatments to yield low molecular weight degradation products, including phenols (see Section 10.3.2).

10.1.3   Pyrolysis and Gasification

Heating wood to temperatures slightly above 100°C initiates some thermal decomposition. A more active decomposition takes place above 250°C, and for industrial applications temperatures up to 500°C may be used. Above 270°C, thermal decomposition does not require any external heat source because the process becomes exothermic. The thermal decomposition of wood is usually called *pyrolysis* or *carbonization*. A number of other terms such as wood distillation, destructive distillation, and dry distillation are used interchangeably for this type of processing.

During the nineteenth century wood distillation was locally practiced to produce various chemicals such as methanol, turpentine, acetic acid, phenols, and wood tar. Together with charcoal and wood these products were important commodities in many communities. Today petrochemicals have completely displaced them and wood pyrolysis is no longer economical. However, in recent times the shortage and high costs of fossil fuels have created new interest in the possibilities of wood pyrolysis, and processes developed for municipal solid waste disposal may also be applied to wood.

In the traditional "wood distillation industry" hardwood was preferred for production of chemicals. Hardwood distillation was formerly an important source for production of acetic acid, methanol, and acetone which were the primary products of this process. The heat required for pyrolysis was generated by burning gas, oil, or coal. In the thermal degradation of wood the volatile components are distillable and can be recovered as liquids after condensation (Fig. 10-2). The solid residue, *charcoal*, is mainly composed of carbon. At higher temperatures the carbon content is increased because of a more complete dehydration and removal of volatile degradation products. Charcoal is mainly used as combustible material for special purposes. A number of charcoal products are known, including activated carbon for adsorption purposes.

A part of the wood tar is soluble in the distillate formed in the process and is recovered during refining. The heavier portion is separated as an insoluble fraction. The settled tars can be fractionated into light oils consisting of aldehydes, ketones, acids, and esters. Various phenols containing a high proportion of cresols and pitch are present in the heavy tar fraction.

The noncondensable gases formed in the distillation of hardwoods usually contain carbon dioxide, carbon monoxide, hydrogen, methane, and other hydrocarbons as the main constituents. These gases can be recycled in the system to product heat, resulting in a lesser need of outside energy and reduced atmospheric pollution.

Destructive distillation of softwoods gives lower yields of acetic acid and

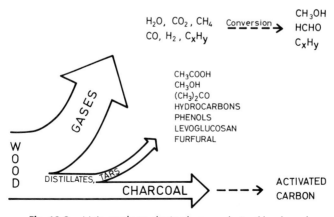

Fig. 10-2. Main products obtained on pyrolysis of hardwood.

methanol than hardwood distillation. Longleaf and slash pine, especially in the form of recovered stumps, were used because of their high resin content. Turpentine and pine oil were the principal products. However, like hardwood distillation, the old softwood distillation process is no longer practiced because high-quality tall oil and turpentine can be produced more economically as by-products of the kraft pulp industry (cf. Section 10.3.1).

Gasification of wood means its thermal degradation by means of partial combustion. For gasification, high temperatures of 1000°C and above are used to produce a gas mixture containing carbon monoxide, hydrogen, methane, and other hydrocarbons. Although gasification has much broader application potential for the conversion of abundant coal supplies to more convenient fuels, the technology is generally applicable to biomass gasification, including wood waste. The process design is under extensive development to maximize the yields of liquid and gaseous fuels. The gas can be used either as a fuel or for the synthesis of such chemicals as methanol or ammonia.

10.2 By-Products after Sulfite Pulping

The majority of the material in sulfite spent liquors originates from lignin (lignosulfonates) and hemicelluloses (Tables 7-5 and 10-1). A variety of useful products can be isolated and prepared from these liquors (Table 10-2)

TABLE 10-1. Products Obtained on Acid Sulfite Pulping^a

Component	Spruce 52% yield	Birch 49% yield
Lignosulfonates^b	480 (540)	370
M > 5000	245 (280)	55
Carbohydrates^c	280	375
Arabinose	10	10
Xylose	60	340
Mannose	120	10
Galactose	50	10
Glucose	40	5
Aldonic acids	50	95
Acetic acid	40	100
Extractives	40	40
Other compounds	40	60

^{a} Approximate values given in kilograms per ton of pulp.
^{b} Calculated as lignin. The estimated values for lignosulfonates are given in parentheses (calculated by assuming the degree of sulfonation to be 2.5 meq/g lignin).
^{c} 80–85% of carbohydrates comprises monosaccharides, the rest is poly- and oligosaccharides.

TABLE 10-2. Main By-Products Based on the Sulfite Spent Liquor and Condensates

Product	Quantity (kg/ton of pulp)	Separation method	Utilization
Lignosulfonates	400–550	Evaporation, precipitation, ultrafiltration, electrodialysis, ion exclusion	Additive (oil well drilling muds, Portland cement concrete), dispersing agent and binder (textiles, products of printing industry, mineral slurries), raw material (vanillin, dimethyl sulfoxide, etc.)
Sugars and aldonic acids	300–450	Evaporation, ultrafiltration, ion exchange, electrodialysis, ion exclusion	Food and chemical industries
Ethyl alcohol	40–60	Fermentation, distillation	Food and chemical industries (solvent, raw material)
Protein	90–110	Fermentation	Food and fodder industries
Acetic acid	30–80	Extraction, distillation	Food and chemical industries (solvent, raw material)
Furfural	10–15	Adsorption, distillation	Plastic and lacquer industries
Butanol, acetone, lactic acid	30–40	Fermentation, extraction, distillation	Plastic and lacquer industries (solvent, raw material)
Cymene	0.3–1	Distillation	Plastic and lacquer industries (solvent)

even if complicated separation methods are often required. The heat value of the lignosulfonates is considerably higher than that of the carbohydrates.

10.2.1   Terpenes

During acid sulfite pulping α-pinene and some monocyclic terpenes are partially converted to p-cymene (Fig. 7-18). Crude cymene can be separated from the digester gas relief condensates and purified by distillation. With spruce and fir the distilled product is 99% p-cymene and only minor quantities of other products, such as borneol and sesquiterpenes, are present. Pine wood contains 3-carene, which affords appreciable quantities of m-cymene besides p-cymene. The crude cymene can be used within the mill as a resin-cleaning solvent. The distilled p-cymene finds use in the paint and varnish industry.

10.2.2   Lignosulfonates

Lignosulfonates are isolated from sulfite spent liquors in more or less pure form depending on the application purposes. They are directly useful because of their dispersion and adhesion properties. For instance, when added to concrete, lignosulfonates are adsorbed on the mineral surface and less water is needed to provide the fluidity and plasticity necessary for handling. This results in a less permeable and stronger set concrete. Similarly, when lignosulfonates are added to mineral slurries and drilling muds, the viscosity is reduced. Lignosulfonates are also used in oil well drilling muds and as binders in animal food pellets. Concentrated sulfite spent liquor also finds widespread use as a dust binder for gravel roads. The use of polymeric lignosulfonates includes a number of other applications as well. One interesting possibility is the preparation of ion exchangers from lignosulfonates. Despite of intensive studies, however, no commercially attractive products have been developed. The resulting products are so far unsatisfactory with respect to the ion exchange capacity and insolubility, etc., and are therefore not competitive with synthetic ion exchange resins.

A variety of methods are available for isolation and purification of lignosulfonates. In the Howard process calcium lignosulfonates are precipitated by adding lime to the spent liquor. Lignosulfonates of improved purity can be isolated after addition of quarternary ammonium salts. In recent years ultrafiltration and ion exclusion techniques have been applied for purification of lignosulfonates. These methods are also useful for further fractionation of the lignosulfonates according to their molecular weight.

Another category of lignin-based chemicals is obtained when lignosulfonates are degraded to low molecular weight products. The commercially most important product is vanillin, which is obtained by alkaline oxidation of softwood lignosulfonates. Hardwood lignin, on the other hand, also gives syringaldehyde because of its high content of syringyl groups and is therefore not suitable for vanillin production.

10.2.3 Carbohydrates and Aliphatic Acids

In principle, the same carbohydrates and their degradation products formed after hydrolysis of wood can be recovered from sulfite spent liquors. However, this requires complicated and expensive separation techniques. The industrial use of sulfite spent liquor components is mainly limited to fermentation processes. The most common product is ethyl alcohol which is produced from hexose sugars by yeast (*Saccharomyces cerevisae*) and separated from the mixture by distillation. Even the carbon dioxide formed in the process can be recovered. Other fermentation products, including acetone, n-butanol, and lactic acid, can be produced by certain microorganisms. Because some contaminants, for example, sulfur dioxide, inhibit the growth of the yeast, they must be removed from the liquor prior to the fermentation.

For overall utilization of the aliphatic components in the sulfite spent liquor, it is advantageous to produce proteins by means of aerobic cultivation using either yeasts (*Candida utilis*) or other fungi (e.g., *Paecilomyces varioti*). In the production of Torula yeast and so-called Pekilo protein not only the hexoses but also pentoses, aldonic acids, and acetic acid are consumed. These cultivation processes are often used for pollution abatement purposes because the biological oxygen demand of the mill effluents is considerably reduced. In the growth of the microorganisms nitrogen and other nutrients are essential. The final protein product contains various amino acids and vitamins and is mainly used as an animal fodder.

Hardwood cooking liquors as well as the resulting evaporation condensates from acid liquors contain appreciable quantities of acetic acid. Because the solutions are very dilute, an economic recovery of the acetic acid is difficult but highly desirable to avoid pollution from discharges. Acetic acid can be recovered from neutral sulfite cooking liquors by extraction of the acidified liquor with organic solvents followed by distillation. Formic acid present in small quantities must be removed by azeotropic distillation if a pure product is required.

In principle, monosaccharides and their conversion products including furfural can be isolated from sulfite spent liquors. Because of the complicated separation technique needed and since alternative raw material sources, such as wood and agricultural wastes are available, these processes have so far been of very limited practical interest. Because of their carbohydrate content, sulfite spent liquors find use either directly or after some fractionation as a feed component for cattle.

10.3   By-Products after Kraft Pulping

Besides lignin degradation products, kraft black liquors contain another large fraction composed of a variety of aliphatic acids derived from polysaccharides (Tables 10-3 and 7-8). A minor but valuable part originates from extractives. The kraft pulp industry has developed advanced techniques for the regeneration of cooking chemicals in which mainly the degradation products of lignin and carbohydrates are burned. Like lignosulfonic acids, kraft lignin has a much higher heat value than the carbohydrates and lower aliphatic acids.

10.3.1   Sulfate Turpentine and Tall Oil

Traditionally the most important by-products of the kraft pulp industry are *sulfate turpentine* and *tall oil** used after refining in the paint and lacquer industry and for a number of other purposes. The yield of these products varies greatly depending on the wood material and especially on the storage of the wood. Crude sulfate turpentine is recovered from the digester relief condensates in a yield of about 10 kg/ton of pulp. The organic sulfur impurities, largely methyl mercaptan and dimethyl sulfide and higher terpenes, are removed in the distillation process. In Scandinavia, the middle fraction, about 75% of the crude turpentine, contains lower terpenes. The distilled product from Scots pine is composed of 50–60% pinene (mostly α-pinene) and 35–40% 3-carene in addition to various monoterpenes and terpenoids present only in minor quantities. The refined fraction originating from southern pines contains considerable quantities of α- and β-pinenes but no 3-carene. If needed, the individual components can be separated and used as such or processed further for special purposes.

During evaporation of the black liquor, the sodium salts of resin and fatty acids together with some neutral resin components and other impurities are separated from the water phase as "soap skimmings." Crude tall oil is recovered from these skimmings after acidification with sulfuric acid. Almost all of the crude tall oil is further purified and fractionated by vacuum distillation to yield a fatty acid fraction and a resin acid fraction (Fig. 10-3). The neutral components are distilled as light oil and components polymerized during distillation remain as pitch residue. The composition of the fractions is largely dependent on the raw material. If hardwood is present besides pine wood, the amount of neutral substances is increased in the resin acid fraction, resulting in lower quality of the product (Table 10-4). Methods have been developed for the separation of the neutral constituents from the resin

---

*The name "tall oil" originates from the Swedish word "tall" which means pine.

TABLE 10-3. Products Obtained on Kraft Pulping[a]

Component	Pine 47% yield	Birch 53% yield
Lignin	510	340
Hydroxy acids	310	240
Glycolic	10	15
Lactic	45	45
2-Hydroxybutanoic	15	65
2,5-Dihydroxypentanoic	10	10
Xyloisosaccharinic	15	45
Glucoisosaccharinic	160	35
Formic acid	70	50
Acetic acid	50	120
Resin and/or fatty acids	75	50
Turpentine	10	—
Miscellaneous (mainly neutral substances)	100	85

[a] Approximate values given in kilograms per ton of pulp.

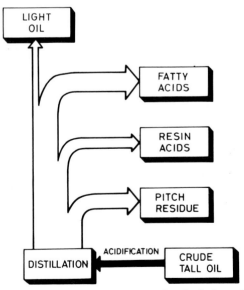

Fig. 10-3. Main distillates obtained from crude tall oil after acidification.

TABLE 10-4. Compositions of Tall (Pine) Oil and the Corresponding Products Originating from Spruce and Birch Wood[a]

Characteristics	Pine	Spruce	Birch
Acid number (mg KOH/g)	160	140	100
Unsaponifiables (%)	7	10	30
Resin acids (%)	40	25	0
Fatty acids (%)	50	60	70

[a] From Kahila (1971).

and fatty acids. An industrially feasible extraction process for this purpose has been developed in Finland. The neutral fraction recovered from the tall soap by extraction contains high amounts of sterols, which can be isolated by crystallization and used as raw material for further processing. Because of the removal of neutral constituents, the quality of the final (distilled) tall oil is improved considerably. Similar extraction processes have been developed in the United States, Canada, and the Soviet Union.

In Scandinavia where Scots pine is the principal wood material in the kraft pulp industry, a normal yield of tall oil in the northern regions is at least 50 kg/ton of pulp but it is considerably lower in the middle or southern regions. In the United States, southern pines also give a tall oil yield of about 50 kg/ton of pulp whereas only about 30 kg/ton can be recovered from Douglas fir in the industry at the Pacific coast.

10.3.2   Kraft Lignin

Sulfate or alkali lignin can be precipitated from black liquors by acidification. The yield of precipitated lignin depends on the pH of the liquor. It is further improved if the black liquor prior to the precipitation is concentrated by evaporation to a higher solids content (25–30%). It is advantageous to use carbon dioxide from stack gases for the neutralization, but the yield is considerably increased if pure carbon dioxide under pressure is applied (Fig. 10-4). Carbon dioxide liberates the phenolic groups with simultaneous formation of sodium hydrogen carbonate. Because of the salting-out effect, the inorganic constituents also contribute to the precipitation. The precipitated lignin can be removed from the solution by filtration, preferably at elevated temperatures (60°–80°C) to improve the filtrability, because the gelatinuous lignin precipitate is aggregated to form a tighter and less hydrated structure. The isolated crude lignin still contains sodium, mainly bound to the carboxyl groups which can be liberated by sulfuric acid at pH 2–3. Addition of sulfuric acid to the liquor after carbon dioxide precipitation results in additional precipitation of lignin at the same time as the carboxylic acids,

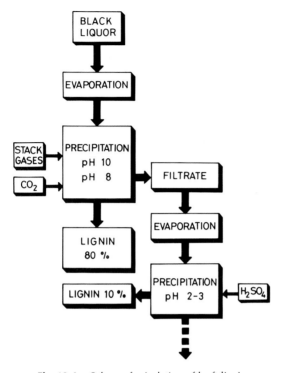

Fig. 10-4.   Scheme for isolation of kraft lignin.

including the hydroxy acids, degradation products from the carbohydrates, are liberated. They may be recovered separately using the procedures described in Section 10.3.3.

Sulfate or alkali lignins find similar uses as lignosulfonates, but their recovery and purification are expensive and at present only marginal quantities are produced. Kraft and alkali lignins can be used as dispersing and stabilizing agents and as additives in rubber, resins, and plastics. Condensation of lignin with formaldehyde and cross-linking with phenols may yield thermosetting polymers useful as adhesives for different products such as paper laminates and plywood. The solubility of the precipitated lignin can be increased by sulfonation. The resulting products compete with lignosulfonates but because of the high content of phenolic hydroxyl groups they are more useful in certain applications, including tannin agents. More definite and narrow fractions with respect to the molecular weight can be obtained by fractionating kraft or alkali lignins by ultrafiltration. Compared to unfractionated lignin, the fractionated product is superior in many applications, for example, for adhesives.

The lignin polymer can also be degraded to low molecular weight products to give a variety of chemicals. Lignin is partly demethylated in the kraft process because of the nucleophilic action of the hydrosulfide ions, resulting in volatile sulfur compounds, mainly methyl mercaptan and dimethyl sulfide (see p. 131 and Fig. 7-28). When concentrated kraft black liquors are heated at 200°–250°C in the presence of additional sulfur, a more extensive demethylation occurs, which affords considerable quantities of dimethyl sulfide (Fig. 10-5). Production of dimethyl sulfide, initiated by Crown-Zellerbach in the United States, does not affect the recovery of sodium and sulfur or lower the heat of combustion. The concentrated black liquor is introduced into a reactor and, after demethylation, the dimethyl sulfide produced and some methyl mercaptan, formed simultaneously, are removed by flashing. They can be used for several purposes, for example as solvents for inorganic salts. Dimethyl sulfide is oxidized by nitrogen tetroxide in the presence of oxygen to dimethyl sulfoxide, which is a very useful product, mainly because of its excellent solvent properties. On further oxidation dimethyl sulfoxide can be converted to dimethyl sulfone.

Numerous attempts have been made to produce other low molecular weight products from lignin by subjecting concentrated kraft black liquors directly to various treatments, including hydrogenation, oxidation, or only heating in the presence of excess alkali.

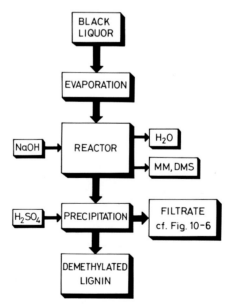

Fig. 10-5. Scheme for production of methylmercaptan (MM), dimethyl sulfide (DMS), and demethylated lignin.

When heating black liquor under pressure in the presence of excess alkali and sulfide at 250°–290°C, lignin is extensively demethylated and partly degraded to low molecular weight phenols and aliphatic acids. Processes based on precipitation and extraction have been proposed for the recovery and fractionation of the reaction products (Figs. 10-5 and 10-6). After demethylation, the reactivity of lignin is increased and it is more easily converted to condensation polymers with certain reagents, such as formaldehyde. With few exceptions no commercial processes have emerged from all this research. One great difficulty is the separation of individual products from the complicated mixture of compounds formed after such treatments.

10.3.3   Carbohydrate Degradation Products

Large quantities of hydroxy acids as well as acetic and formic acids are formed during kraft pulping (Tables 10-3 and 7-8). Among hydroxy acids from softwood, glucoisosaccharinic acid predominates. Especially when considering that the heat value of these hydroxy acids is only 25–50% of that of kraft lignin, their recovery seems motivated, but adequate markets have not yet been established for these products. Liberation of the aliphatic acids

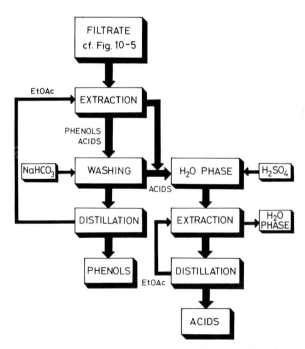

Fig. 10-6.   Separation of low molecular weight phenols and aliphatic acids.

before isolation would also result in radical changes in the sodium and sulfur balance and recovery. Various methods have, however, been proposed for the isolation of these aliphatic acids, including extraction, ultrafiltration, and distillation under reduced pressure. Many of them, including formic, acetic, and lactic acid are well-known chemicals. The potential market for other acids, such as isosaccharinic acids, would include applications in which products having sequestering properties are needed. They could also be used as starting materials for production of a variety of chemicals and polymers.

References

Alén, R., Patja, P., and Sjöström, E. (1979). Carbon dioxide precipitation of lignin from pine kraft black liquor. *Tappi* **62**(11), 108–110.

Ander, P., and Eriksson, K.-E. (1978). Lignin degradation and utilization by micro-organisms. *Prog. Ind. Microbiol.* **14**, 1–58.

Andersen, R. F. (1979). Production of food yeast from spent sulphite liquor. *Pulp Pap. Can.* **80**(4), 43–45.

Bansal, I. K., and Wiley, A. J. (1974). Fractionation of spent sulphite liquor using ultrafiltration cellulose acetate membranes. *Environ. Sci. Technol.* **8**, 1085–1090.

Bansal, I. K., and Wiley, A. J. (1975). Membrane processes for fractionation and concentration of spent sulfite liquors. *Tappi* **58**(1), 125–130.

Bar-Sinai, Y. L., and Wayman, M. (1976). Separation of sugars and lignin in spent sulfite liquor by hydrolysis and ultrafiltration. *Tappi* **59**(3), 112–114.

Bicho, J. G., Zavarin, E., and Brink, D. L. (1966). Oxidative degradation of wood. II. Products of alkaline nitrobenzene oxidation by a methylation-gas chromatographic technique. *Tappi* **49**, 218–226.

Bratt, L. C. (1979). Wood-derived chemicals: Trends in production in the U.S. *Pulp Pap.* **53**(6), 102–108.

Collins, J. W., Boggs, L. A., Webb, A. A., and Wiley, A. A. (1973). Spent sulfite liquor reducing sugar purification by ultrafiltration with dynamic membranes. *Tappi* **56**(6), 121–124.

Compere, A. L., and Griffith, W. L. (1980). Industrial chemicals and chemical feedstocks from wood pulping waste waters. *Tappi* **63**(2), 101–104.

Counsell, J. N., ed. (1977). "Xylitol." Applied Science Publ., London.

Eriksson, K.-E. (1980). Development of biotechnology within the pulp and paper industry. *Int. Congr. Pure Appl. Chem. (IUPAC), 27th, Helsinki*, pp. 331–337. Pergamon, Oxford.

Esser, M. H., ed. (1979). *Annu. Lightwood Res. Conf. Proc., 6th, Atlanta, Ga.* Southeast. For. Exp. Stn., Asheville, North Carolina.

Forss, K. (1974). Possibilities of developing chemical products from spent sulphite liquors. *Pap. Puu* **56**, 174–178.

Forss, K., and Fuhrmann, A. (1976). KARATEX—the lignin-based adhesive for plywood, particle board and fibre board. *Pap. Puu* **58**, 817–824.

Forss, K., and Passinen, K. (1976). Utilization of the spent sulphite liquor components in the Pekilo protein process and the influence of the process upon the environmental problems of a sulphite mill. *Pap. Puu* **58**, 608–618.

Frank, E., Hirschberg, H. G., and Pfeiffer, H. J. (1976). Hydrolysis of natural fibrous materials. *"Achema 76"* Spec. No., pp. 1–11.

Goheen, D. W. (1971). Low molecular weight chemicals. In "Lignins" (K. V. Sarkanen and C. H. Ludwig, eds.), pp. 797-831. Wiley (Interscience), New York.

Goldstein, I. S. (1980). New technology for new uses of wood. Tappi **63**(2), 105-108.

Hergert, H. L., Van Blaricom, L. E., Steinberg, J. L., and Gray, K. R. (1965). Isolation and properties of dispersants from Western Hemlock bark. For. Prod. J. **15**, 485-491.

Herric, F. W., and Hergert, H. L. (1977). Utilization of chemicals from wood: retrospect and prospect. The structure, biosynthesis and degradation of wood. Recent Adv. Phytochem. **2**, 443-515.

Holmbom, B. (1978). Constituents of tall oil. A study of tall oil processes and products. Ph.D. Thesis, Univ. of Åbo, Åbo, Finland.

Hoyt, C. H., and Goheen, D. W. (1971). Polymeric products. In "Lignins" (K. V. Sarkanen and C H. Ludwig, eds.), pp. 833-865. Wiley (Interscience), New York.

Ivermark, R., and Jansson, H. (1970). Recovery of tall oil from kraft pulp mills. Sven. Papperstidn. **73**, 97-102. (In Swed.)

Kahila, S. K. (1971). Yield, quality and composition of crude tall oil. Kem. Teol. **28**, 745-756. (In Finn.)

Kent, J. A., ed. (1974). "Riegel's Handbook of Industrial Chemistry," 7th ed. Van Nostrand-Reinhold, New York.

Kringstad, K. (1977). "Forest Industry By-Products as Raw Material for Chemicals and Proteins," Rep. No. 65. Swed. Agency Tech. Dev., Stockholm. (In Swed.)

Kringstad, K. (1980). The challenge of lignin. In "Chemrawn I," IUPAC, p. 627. Pergamon, Oxford.

Maloney, G. T. (1978). "Chemicals from Pulp and Wood Waste. Production and Applications." Noyes Data Corp., Park Ridge, New Jersey.

Perret, J.-M., Garceau, J. J., Pineault, G., and Lo, S.-N. (1976). Treatment of spent bisulfite liquor by the technique of ion-exclusion. Pulp Pap. Can. **77**(11), 107-110.

Rogers, I. H., Manville, J. R., and Sahota, T. (1974). Juvenile hormone analogs in conifers. II. Isolation, identification, and biological activity of cis-4-[1'(R)-5'-dimethyl-3'-oxohexyl]-cyclohexane-1-carboxylic acid and (+)-4(R)-5'-dimethyl-3'-oxohexyl]-1-cyclohexene-1-carboxylic acid from Douglas-fir wood. Can. J. Chem. **52**, 1192-1199.

Rychtera, M., Barta, J., Fiecter, A., and Einsele, A. A. (1977). Several aspects of the yeast cultivation on sulphite waste liquors and synthetic ethanol. Process Biochem. **12**(2), 26-30.

Rydholm, S. A. (1965). "Pulping Processes." Wiley (Interscience), New York.

Seidl, R. J. (1980). Energy from wood: A new dimension in utilization. Tappi **63**, 26-29.

Simard, R. E., and Cameron, A. (1974). Fermentation of spent sulphite liquor by Candida utilis. Pulp Pap. Can. **75**, Convention Issue, 107-110.

Slama, K., Romaňuk, M., and Sörm, F. (1973). "Insect Hormones and Bioanalogues," pp. 90-275. Springer-Verlag, Berlin and New York.

Soltes, E. I. (1980). Pyrolysis of wood residues. A route to chemical and energy products for the forest products industry? Tappi **63**(7), 75-77.

Tillman, D. A. (1978). "Wood as an Energy Resource." Academic Press, New York.

Timell, T., ed. (1975). "Wood Chemicals—A Future Challenge," Vol. 1, Applied Polymer Symposium, No. 28. Wiley (Interscience), New York.

Weiss, D. E. (1979). Energy from biomass. Appita **33**, 101-110.

Wiley, A. J., Scharpf, K., Bansal, I., and Arps, D. (1972). Reverse osmosis concentration of spent liquor solids in press liquors from high-density pulps. Tappi **55**, 1671-1675.

APPENDIX

Chemical Composition of Various Wood Species[a]

Species	Common name	Total extractives	Lignin	Cellulose	Glucomannan[b]	Glucuronoxylan[c]	Other polysaccharides	Residual constituents
Softwoods								
Abies balsamea	Balsam fir	2.7	29.1	38.8	17.4	8.4	2.7	0.9
Pseudotsuga menziesii	Douglas fir	5.3	29.3	38.8	17.5	5.4	3.4	0.0
Tsuga canadensis	Eastern hemlock	3.4	30.5	37.7	18.5	6.5	2.9	0.5
Juniperus communis	Common juniper	3.2	32.1	33.0	16.4	10.7	3.2	1.4
Pinus radiata	Monterey pine	1.8	27.2	37.4	20.4	8.5	4.3	0.4
Pinus sylvestris	Scots pine	3.5	27.7	40.0	16.0	8.9	3.6	0.3
Picea abies	Norway spruce	1.7	27.4	41.7	16.3	8.6	3.4	0.9
Picea glauca	White spruce	2.1	27.5	39.5	17.2	10.4	3.0	0.3
Larix sibirica	Siberian larch	1.8	26.8	41.4	14.1	6.8	8.7	0.4
Hardwoods								
Acer rubrum	Red maple	3.2	25.4	42.0	3.1	22.1	3.7	0.5
Acer saccharum	Sugar maple	2.5	25.2	40.7	3.7	23.6	3.5	0.8
Fagus sylvatica	Common beech	1.2	24.8	39.4	1.3	27.8	4.2	1.3
Betula verrucosa	Silver birch	3.2	22.0	41.0	2.3	27.5	2.6	1.4
Betula papyrifera	Paper birch	2.6	21.4	39.4	1.4	29.7	3.4	2.1
Alnus incana	Gray alder	4.6	24.8	38.3	2.8	25.8	2.3	1.4
Eucalyptus camaldulensis	River red gum	2.8	31.3	45.0	3.1	14.1	2.0	1.7
Eucalyptus globulus	Blue gum	1.3	21.9	51.3	1.4	19.9	3.9	0.3
Gmelina arborea	Yemane	4.6	26.1	47.3	3.2	15.4	2.5	0.9
Acacia mollissima	Black wattle	1.8	20.8	42.9	2.6	28.2	2.8	0.9
Ochroma lagopus	Balsa	2.0	21.5	47.7	3.0	21.7	2.9	1.2

[a] J. Janson, P. Haglund, and E. Sjöström, unpublished data. All values are given as % of the dry wood weight.
[b] Including galactose and acetyl in softwood.
[c] Including arabinose in softwood and acetyl in hardwood.

INDEX